25 WALKS

FIFE

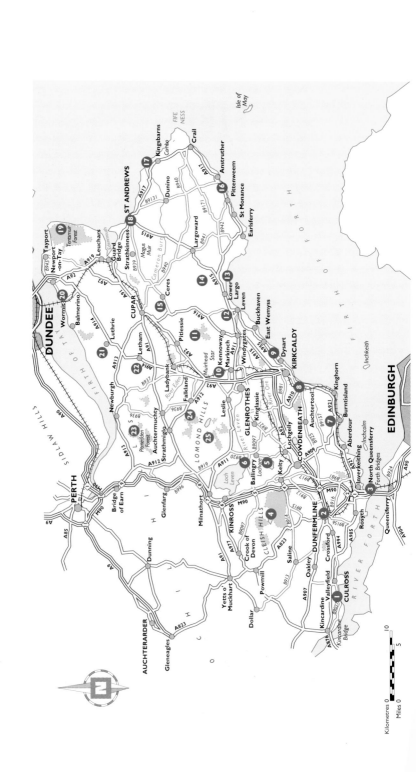

25 WALKS

FIFE

Hamish Brown

Series Editor: Roger Smith

MERCAT PRESS
www.mercatpress.com

Acknowledgements

Many people gave help and advice in the compilation of this selection of walks. I would especially like to thank Fife Ranger Services; Cathy Kinnear (Fife Council); Ken Shaw (RSPB); Richard and Edith Cormack; Dorothy Wilson; Val and Ian Brown; John Mitchell; Clare White and other NTS staff; Scottish National Heritage staff; Forest Enterprise; and the many farmers and local walkers who were invariably proud of their Fife heritage.

First published, 1995

Revised edition published, 2001

Published by
Mercat Press
53 South Bridge
Edinburgh
EH1 1YS
www.mercatpress.com

ISBN: 1 841830 089

Cover illustration: Crail Harbour
All photographs © Hamish Brown

Printed in China through World Print Ltd.

CONTENTS

INTRODUCTION

Fife is a 'Kingdom' which, once discovered, will always delight: a green and varied landscape, a coastline which lives up to its royal description as 'fringed with gold', hills with views out of all proportion to their altitude, fascinating historic towns, castles and prehistoric sites – Fife has it all.

There is plenty of variety in the length and physical demands of this selection of Fife walks but most are well within the capabilities of ordinary walkers rather than any super-fit Munro-bagger. Many are circular routes with shorter versions available, linear routes on the coast have bus/train services to return to the start, and many routes would be enjoyed on family outings – something for everyone – but do allow plenty of time as there is much to see and do.

Several of the coastal walks are described more fully in my book *The Fife Coast* (Mainstream, 1994), the Regional Council has issued good books on Fife's castles and archaeological heritage, and both the Buildings of Scotland (J. Gifford) and RIAS (G. Pride) series have superb illustrated guides to buildings and monuments and challenging days. A national award scheme in 1993 named Fife's Regional Council as the most environmentally friendly in Scotland. Several walks are in the Fife Regional Park, several take in National Trust for Scotland properties, while some feel surprisingly remote, and both the Crail–St Andrews coast walk and the round of the Lomonds are full and challenging days. An official Fife Coastal Path is steadily being developed, and on some walks you will see its distinctive signposting. Fife Council also operate a public transport information line (weekdays, 0900–1600, tel: 01592 416060).

The top visitor venues in Fife are the two sea centres, the Scottish Deer Centre and the Kirkcaldy Museum and Art Gallery. Three of these are on walks and the Deer Centre, on the A91 near Cupar, can easily be added after any of the walks in central or north Fife. Attractions near walks are briefly indicated. It must be confessed that a few walks on the edge of Fife stray into Kinross-shire or Perthshire, and one walk enters from the Lothians. I apologise for this poaching!

Unlike walking in the Highlands, many of the walks are excellent in winter when cold, crisp days can make walking a delight. Some consider a place like Vane Farm best in winter because of its seasonal influx of birds, which could also apply to the coast. Fife is a surprisingly hilly county but sea or lochs are constantly in view. Spring can be vibrantly green, early-summer rich and the autumn colours magnificent, while July and August often disappoint. June is my favourite month: most places to visit are open, the landscape is patched with yellows of oilseed rape and whin, the stirring skylarks reel over the hills. Welcome to the Kingdom of Fife.

USEFUL INFORMATION

The length of each walk is given in kilometres and miles, but within the text measurements are metric for simplicity. The walks are described in detail and are supported by accompanying maps (study them before you start the walk), so there is little likelihood of getting lost, but if you want a back-up you will find the 1:50 000 Landranger Ordnance Survey maps on sale locally. New 1:25 000 Explorer maps are being developed.

Every care has been taken to make the descriptions and maps as accurate as possible, but the author and publishers can accept no responsibility for errors, however caused. The countryside is always changing and there will inevitably be alterations to some aspects of these walks as time goes by. The publishers and author would be happy to receive comments and suggested alterations for future editions of the book.

In the text the word 'path' is used for a purely pedestrian route, 'track' for private, unsurfaced but motorable routes (such as farm and forestry tracks) and 'road' for a tarred, public way.

A reminder about country behaviour: farm gates must be closed when used, farm buildings and machinery left alone, dogs kept on leads near livestock, growing crops avoided, matches not tossed aside or litter dropped. Routes described may or may not be rights-of-way. Farmers are usually friendly folk who work long hours to maintain the land we enjoy. When you meet, be helpful and friendly too!

The weather in Fife can be changeable but enjoys an east-coast freshness and relatively low rainfall. If cloud covers the hills, walks lower down or on the coast can be rewarding. When you may be out for several hours then water/windproof garments and strong footwear such as boots are advised, in most cases elsewhere shoes/trainers would suffice.

Warning sign – a particular problem in Fife.

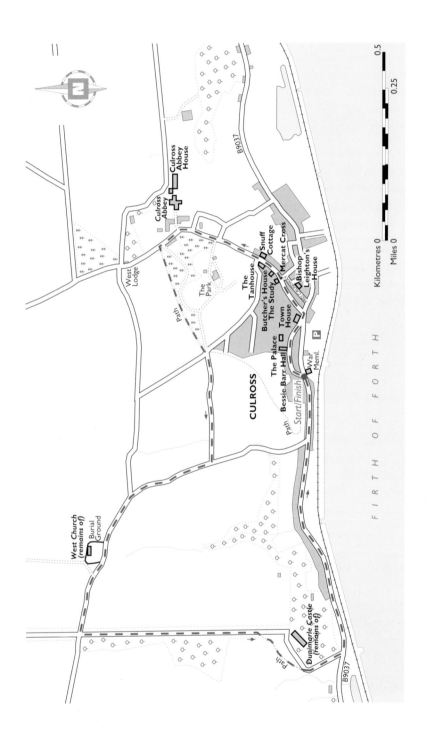

HISTORIC CULROSS

Culross, a small town lying on the Forth Estuary, is a unique 17th-century survival and well worth exploring – a sanitised step back and walk in the period when stone houses were first developing. The narrow cobbled streets and the houses with their red pantiles and crowstep gables have often been used as a film set. Early on in the walk you pass the Town House (NTS) where a short historical video and guidebooks can fill out this brief introduction.

From the time of the monks, coal was a major industry, and cheap fuel beside the sea brought salt panning to Culross. (In 1633 there were 50 pans, so the town must have been a smoky hole.) Ships returning from Holland brought the red pantiles as ballast. Some of the first NTS purchases in the 1930s were made here: the Study cost £150, nine other houses, £168. Longannet power station, prominent upriver, is still coal-fired.

Start at the carpark, which has display boards, anchors and the curved shape of a boathouse, as well as a toilet, war memorial and children's play area. Then cut down the lane immediately past the building close to the road, which has a lintel date of 1681, and turn right for two NTS showpieces. Bessie Bar's malthouse (tearoom) is the building with the forestair up its gable, which is unusual, and, next to it, is the gem of the ochre-washed Palace, built by Sir George Bruce, part in 1597, the rest in 1611. It has rare tempera-painted ceilings and period furnishings; note the windows with their mix of glass and shutters. The Town House (tolbooth), with its clock tower, shows Dutch influences. In 1975, when town councils ceased to exist, it was handed over to the NTS as an interpretative centre.

Turn up the next street, the Back Causeway, cobbled and with a clear 'croon o' the caussey' (crown of the

INFORMATION

Distance: 3½km (2½miles).

Start and finish: Main carpark, west end of Culross, on B9037 between Forth Bridge and Kincardine Bridge.

Terrain: Pavements and paths; no special footwear needed.

Toilets: At the carpark and some sites, cafés.

Refreshments: Bessie Bar's malthouse (NTS) and two pubs.

Opening hours: *Palace:* Easter, May–September, 1100–1700. *Town House:* 1100–1300, 1400–1700. *Study:* 1300–1600. Tel: 01383 880359.

Culross Abbey ruins and church.

Houses on Back Causeway.

causeway), the larger stones where the gentry walked clear of the street's rubbish. The narrow tower ahead is the Study, and the corbelled room at the top is thought to be where the 17th-century Bishop Leighton worked. Across the street a date on a gable, 1577, is the earliest in the town. The mercat cross goes back to 1588. A merchant's house two up from the Study has a Greek inscription that translates: 'God provides and will provide.' Before going up the Tanhouse Brae, do walk down the Mid Causeway to see Bishop Leighton's house.

On the Tanhouse Brae, left, a butcher's house has a cleaver and steel portrayed and, further on, on the right, the Snuff Cottage (1673) with a squeezed-in window, has the motto 'Wha wad ha thocht it' (who would have thought it?). This is in fact the first line of a couplet, the second line – 'Noses ha' bocht it' (noses have bought it) – appearing on the lintel of the snuff merchant's premises in Edinburgh. The Tanhouse is on the corner, but you keep on up the Brae, aiming for the castle-like tower of the church, below which are some of the abbey ruins.

A church has stood here since early times. St Serf was here in the 5th century and a pupil of his – traditionally found as a newborn babe on the shore with his mother – was to become St Mungo, the patron saint of Glasgow. An abbey was founded in 1217 but fell into decay after the Reformation. The remoter West Church was the parish church then, but the congregation successfully petitioned parliament in 1633 and the abbey became the church; the present building being the original choir. A notable feature is the Bruce Aisle, off the north transept, where the family burial place has marble parental figures and a kneeling row of three sons and six daughters, all in period dress.

Outside, turn right to circle the graveyard. At a gate Culross Abbey House is visible. It dates from 1608, very little later than the Palace but years ahead in style and gracious living (not open to the public). Cochrane, 10th Earl of Dundonald, and on a par with Nelson as a naval genius, once lived here; he was a figure largely ostracised by the establishment in Britain.

Trade Stones abound in the Culross churchyard: the hammermen's with a crown above the hammer, a butcher's with his cleavers, a gardener's with crossed spade and rake, several with maltsters' paddles and many with the sock and coulter of ploughmen.

Once out of the churchyard, continue up the road leading inland. Note the neat house on the right with its 'eyes' of windows. Just past some lodge gates follow the footpath sign for the West Kirk. Cross a minor road, continue on the track and then turn right at a T-junction and finally left at a Y-fork. At the West Kirk some of the lintels in the church are old gravestones with incised swords on them. There are many hammermen's stones and a big '4' is the symbol of a merchant/trader.

Ploughman's symbols in Culross churchyard.

Continue along the track till it crosses the tree-lined but abandoned driveway of Donimarle Castle. Go through the gate, left, and down this grassy avenue then, just before the last two big trees, veer right on a green path (skirting the castle and its entrance) to come out on the tarred drive which descends to the coast road. Cross this towards the sea and find a footpath beside the railway which leads back to Culross.

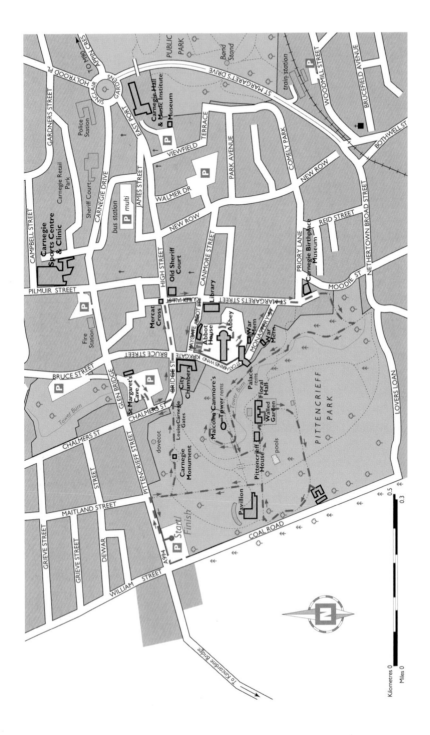

ROUND DUNFERMLINE TOWN

Dunfermline, ancient capital of Scotland, has one of Britain's finest parks, so a walk here gives a strange contrast of past and present, urban and rural, historic and modern. Free heritage walks take place every Saturday and Sunday from April to September; for details, telephone: 01383 724302.

Enter Pittencrieff Park (locally called 'The Glen') from the gates on Pittencrieff Street then turn left at the first opportunity to pass the statue of Andrew Carnegie. Son of a weaver, he was banned from the Pittencrieff estate grounds as a boy, but, later, as the great American industrialist, he bought the estate and promptly gifted it to the town. The huge gates beyond are a memorial to his wife. (To the left there's a small, battlemented doocot [dovecot].)

INFORMATION

Distance: 3km (2 miles).

Start and finish: Carpark, north-west corner of Pittencrieff Park (A994 exit from town).

Terrain: Town, park paths, no special footwear needed.

Toilets: Numerous.

Refreshments: Wide selection in the town and also the Pavilion in the park in summer.

Opening hours: *Cave:* daily, April–September, 1100–1600. *Abbey and Palace Visitor Centre (Historic Scotland):* April–September, daily, 0930–1830; October–March, Monday/Wednesday/Saturday, 0930–1630; Thursday, 0930–1230; Sunday, 1400–1630. *Abbey Parish Church (Bruce's grave):* April–September, Monday–Saturday, 1000–1630; Sunday, 1400–1630. *Pittencrieff House Museum:* Open all year, 1100–1600 (to 1700, April–September). *Pittencrief Park, Floral Hall and Aviary:* open daylight hours. *Carnegie Birthplace Museum:* April–October, Monday–Saturday, 1100–1700; Sunday, 1400–1700; November–March, daily, 1400–1600. *Tourist Information:* 13 Maygate, April–September, 01383 720999.

In the formal garden of the Glen.

The City Chambers from the abbey grounds.

Go through the gates and turn left then cross and enter the carpark to see St Margaret's Cave. It is actually *under* the carpark. Margaret was an English princess, born in exile, who returned to the court of Edward the Confessor when she was eight. There she first met Malcolm, a refugee from Scotland, whose father, King Duncan, had been murdered by Macbeth. Fleeing again, when 20, her ship was driven into the Forth and she was helped by Malcolm, who then married her (1070). Devout Margaret had such an influence on Scotland that she was canonised in 1249.

Walk up the High Street, which becomes pedestrianised and has the old mercat cross facing the tall spire of what was built as a guildhall. The French baroque château-style building with the clock tower is the City Chambers.

Turn down Guildhall Street and right at the first crossroads. The impressive Central Library building is a Carnegie gift to the town. The tourist office is worth a visit for a town map or any queries before visiting the Abbot House next door. This, the oldest building in the town, has been restored, and was opened in 1995 as a Heritage Centre.

Continue along the Maygate until a gate gives access to the abbey grounds. Have a look at some of the 'trade' gravestones as you wander round leftwards, which leads you to the end of the abbey and the site of St Margaret's shrine (only the marble base survives).

When you have almost completed the circuit, you will see the visitor centre over on the left, which is housed in the shell of what was a partly monastic and partly a palace building (the unfortunate Charles I was born here). After a look at the west front, enter the abbey with its impressive Norman pillars and continue into the parish church, which was rebuilt in the nineteenth century on the ruined choir – the wording 'King Robert the Bruce' around the abbey tower was added at this time.

The King's final resting-place is under the pulpit. When discovered, it was noted that his ribcage had been sawn open to remove the heart, which had been carried on a crusade to Spain, and is now buried at Melrose Abbey. There is a bronze cast of his skull on display.

Carnegie's birthplace.

Leave by the south perimeter wall, left (east), with fine views to the Forth, to walk down steps and go right to pass the 1939–1945 war memorial on to Monastery Street. Walk past the 1914–1918 war memorial opposite and turn right to reach the Andrew Carnegie birthplace museum, a modest cottage (extended) with some interesting exhibits.

Down and across, a gate leads into the Glen. Swing right before the play area to follow the meanderings of the Tower Burn. Keep beside it even when led through *under* the double-arched bridge. At a small footbridge over the burn, however, turn off right, to twist up and out on to a road.

At one time, this was the main town access from the west; hence the bridge on top of a bridge as the steep climb was modified. Immediately right is a Historic Scotland notice about Malcolm Canmore's tower, which you can now visit – though only foundations survive. Walk to, and past, the tower to descend to the squirrel-haunted double bridge. Peer over then walk on, under a bridge, to come out near the Pavilion. The ground falls away and becomes more traditional parkland and is very attractive. (Tearoom and toilets available.)

Turn left for Pittencrieff House, built about 1610 and renovated early this century by Sir Robert Lorimer. It houses exhibitions. Beyond it are the glasshouses (Floral Hall) with the superbly kept formal gardens lying in front. West, right across the park, is the aviary and animal house. Peacocks are everywhere.

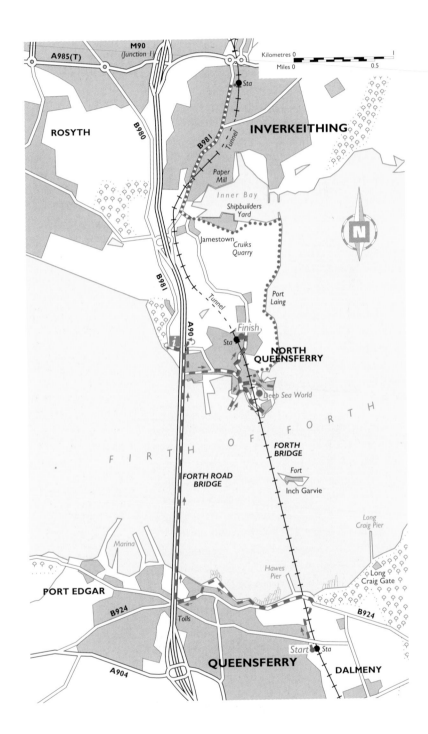

Kilometres 0   1
Miles 0   0.5

A985(T)

M90
(Junction 1)

Sta

ROSYTH

B980

B981

INVERKEITHING

Tunnel

Paper
Mill

Inner Bay

Shipbuilders
Yard

B981

Jamestown

Cruiks
Quarry

Port
Laing

A90

Tunnel

Finish

Sta

NORTH
QUEENSFERRY

Deep Sea World

FIRTH   O F   FORTH

FORTH
BRIDGE

Fort

FORTH ROAD
BRIDGE

Inch Garvie

Long
Craig Pier

Marina

Long
Craig Gate

PORT EDGAR

Hawes
Pier

B924

B924

Tolls

Start Sta

A904

QUEENSFERRY

DALMENY

THE FORTH BRIDGE

This is a romantic way to enter Fife – walking across the Forth Road Bridge and, using the fabulous railway bridge, there is no difficulty returning to the start. Cars can be parked at Dalmeny Station, which lies above Queensferry.

Coming out from the station or its carpark, cross the road to take a footpath running parallel with the railway towards the Forth. Cross an old railway line by a footbridge then head half-right to drop down 120 steps (Jacob's Ladder) to the shore road. Turn left and walk along through Queensferry, a town of considerable charm and antiquity.

Sailings are made from Hawes Pier to Inchcolm (off Aberdour) with its historic abbey, and might appeal. There are several cafés in the town, and much of historical interest to see. The Hawes Inn is the setting of the opening scene in Robert Louis Stevenson's story *Kidnapped* and, until the road bridge was built, was where cars crossed the Forth by ferry. Note the sculpture of a seal and her pup.

The High Street (much of it 16th century) has houses on two levels (as in Chester), while Black Castle is dated 1626 and has a secret stair. The old churchyard has some finely carved stones, one of a full-rigged ship, and many carvings of skulls. The Museum is interesting, and the dominant tower along the High Street is the tolbooth

INFORMATION

Distance: 6km (4 miles) or, to Inverkeithing, 10km (6 miles).

Start and finish: Queensferry (Dalmeny Station); North Queensferry or Inverkeithing.

Terrain: Roads, bridge footway, coastal path – no special footwear required.

Toilets: In Queensferry, North Queensferry and Inverkeithing, also at Visitor Centre at north end of the bridge.

Public transport: Frequent trains to Dalmeny from Edinburgh and Fife stations, and return trains from North Queensferry or Inverkeithing. Enquiries: 01592 416060. Check return train times at Dalmeny at the start of the walk.

Refreshments: Cafés, hotels and restaurants in the same settings.

Opening hours: *Inverkeithing Museum:* All year, Thursday–Sunday, 1200–1600. *Tourist Information Office (Queensferry Lodge Hotel, north end of Forth Road Bridge):* tel: 01383 417759 (bridges exhibition). *Deep Sea World:* All year, daily, 1000–1800.

A stone in the old Queensferry graveyard.

with a commemorative well below it. There are covenanting memories, an old harbour, a friary building and, on the left, a house with marriage lintels, all of which give a flavour of past days. The road bridge now looms ahead; climb the steps to gain this gateway into Fife.

The Forth Bridge from Queensferry.

Head up, passing Plewlandscroft, to reach the bridge. Take the path signed 'Cyclists only' (under the bridge) then turn left up the road to reach the complex of buildings servicing the bridge. No tolls or queuing for walkers! It is 2km to reach the other side and, in gales, you can feel the whole bridge flexing. The towers are 160m in height. The bridge has the lighting, drainage, road area (and servicing and finances) of a small town, and is a memorable walk, however blasé we are about it as motorists. A plaque at the far side marks its opening in 1964 by Queen Elizabeth II, who then made the last ferry sailing. The original ferry crossing dates back to the time of Queen Margaret, wife of Malcolm Canmore, who gave her name to Queensferry.

Turn down the pedestrian steps. By passing under the bridge and up the other side you reach the Visitor Centre, Tourist Information Office, exhibition and hotel/restaurant area, which is a worthwhile diversion. Wend back and down into North Queensferry thereafter. There is an odd cluster of wells and, behind them, the start of the coastal path. (The Napoleon's hat-shaped

The Friary, Inverkeithing.

well aptly commemorates the Battle of Waterloo, 1815.) The coastal path leads round to Inverkeithing (1¼ hours), an optional extra, from where a train can be taken back to the start.

It is worth wandering down to the pier and then round the bay to a viewing area right under the railway bridge – still one of the world's greatest engineering spectacles, the centenary of which, in 1990, was a great occasion. The bridge is floodlit in the evening. Inchgarvie is the island offshore, full of old wartime defences.

On the way back into town turn right at a footpath sign (Helen Place and St James' Chapel) and walk on to reach the Deep Sea World complex – one of Europe's biggest aquariums. The million-gallon tank is the old flooded Battery Quarry and you travel through it along a moving walkway in a transparent tunnel . . . quite an experience! (Stone from the quarry was used on bridge foundations, canals, London streets and a Russian fort.)

The Forth Road Bridge.

Options now include a walk up the steep, twisting road, from the wells, to reach the station where you can catch a train back to the start, or you can take the signed coastal path to Inverkeithing, returning from there by train; you could even sail to Inchcolm Island from Queensferry (*Maid of the Forth*), which is around about a three-hour excursion (for information call 0131 331 4857). The bridge is now officially a national monument, and has been given a 'listed' status.

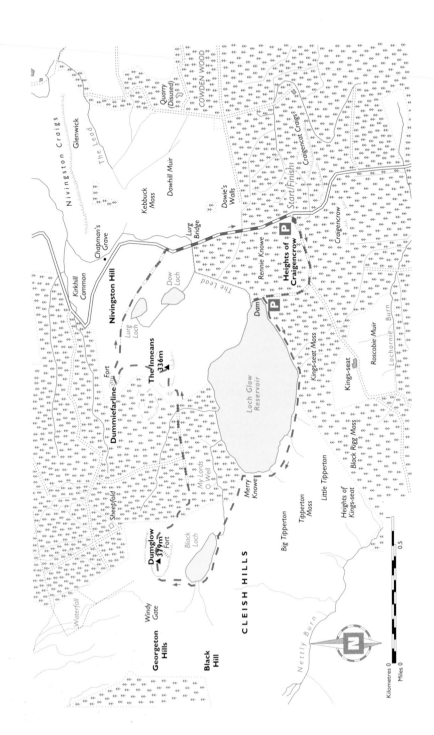

CLEISH HILLS

Georgeton Hills

Black Hill

LOCH GLOW AND THE CLEISH HILLS

T his is a hidden corner of upland country that has been progressively taken over by blanket afforestation, so the open spaciousness of the view from the 379m summit comes as quite a surprise.

The start is a very small off-road carpark, which can be easily missed (see Information panel). There's an East of Scotland Water sign (Customer Services, tel: 01345 420420) – Loch Glow and the Black Loch are particularly popular with anglers. There are actually four lochs hidden away like navels in these bellyfolds of hills: Loch Glow, the Black Loch, Lurg Loch and Dow Loch; the walk circles them all.

There is a locked gate on the track to Loch Glow and a seasonal water-bailiff's caravan. Loch Glow comes as quite a surprise and, if there is nobody about, can appear far bigger than it really is. The walk along the south shore is best made next to the water – the path along the bank above is much boggier. A few grouse survive on the moors but reeds and grass dominate over blaeberry and heather.

Sweep right round the loch to reach a wall at the far corner (Merry Knowe) then turn left to follow its line (which is also the old county boundary). There is a pretty view from a rise with the Black Loch

INFORMATION

Distance: About 8km (5 miles).

Start and finish: Small carpark (GR: 099954), where a forestry track leaves a minor road. Leave the M90 at Junction 4 and head west, turning right after 2½km on to a narrow road, sign-posted to Cleish. The start is on the left, 3km on. (Loch Glow sign at carpark.) When leaving continue along this road and down by Nivingston then turn right to regain the M90 at Junction 5.

Terrain: Mostly boggy moorland or grassy hills. Also several walls/fences and locked gates to be crossed. Waterproof footwear advised.

Refreshments: None en route, but Nivingston Country Hotel Restaurant lies 2½km to north. Tel: (01577) 850216.

Dumglow beyond Loch Glow.

backed by the grassy Georgeton Hills, all with the complex knobbly character of the area's volcanic origins. The south side of the Black Loch is very wet to start with, but thereafter the going is firm and pleasant.

Cross the wall/fence at the end of the loch and start to climb up to the left (west) skyline of Dumglow. The saddle is called Windy Gate, often an apt description. Cross a ruined wall and go up by the fence. Aldie Castle is a landmark on the plain below. This ascent has a very steep section to reach Dumglow's summit trig point at 379m. The view is quite fantastic with the Ochil scarp and Devon Valley leading the eye beyond the Wallace Monument to Ben Lomond; the Fife Lomonds rise beyond Loch Leven; Largo Law is seen over the shoulder of Benarty; the Bass Rock marks the far south-east; and Edinburgh is backed by the Pentland Hills – an astonishing panorama. The trig point sits on a prehistoric mound (a tree-coffin burial) which is circled by the walls of a fort. From the north rim you can look down on Cleish Castle, another typical tower house.

Head off eastwards (across several defensive wall lines) till the slope begins to fall away into forest. Ahead are the triple bumps of The Inneans, your next place to visit. Turn right to follow the fence until a basic stile is reached, cross and follow the path beyond. Over a brow this then leads down into the forest with a big break clearly seen going straight across to the three knobbly hills. Walk this break, cross the fence carefully and climb the central of the three bumps, at 336m almost as fine a viewpoint as Dumglow. The Lurg and Dow Lochs lie below your perch, the last of the day's collection.

Due north, just right of the plantings, which go right over

The volcanic knobbliness of the Loch Glow Hills.

the northern top of The Inneans, you can see another prominent knoll with a big triangular cairn on top. Walk over to this, picking up and using sheep paths. Dummiefarline (The Dummie, locally) is also a prehistoric defensive site. From it you see the road making its way down to Cleish; the parish church is prominent but the castle is hidden.

Head east, picking up a path outside the Lurg Loch bogs and flanking Nivingston Hill. The path evolves into a track of sorts leading to a gate. Continue, bearing right a bit, to come out to the minor road at a gate/small sheepfold. A 10-minute tramp along the road leads back to the start.

If forced to park by Loch Glow then after leaving Dummiefarline, traverse The Inneans and keep to the drier, higher ground back to the dam and carpark.

If you're a fisherman, Loch Glow offers trout fishing from 15 March–6 October; with permits available from the on-site seasonal caravan. If you're into challenging walks you could try the day's outing made by J.H.B. Bell and W. Omand in October 1931: Falkland (10.00), E. Lomond to W. Lomond (11.51), W. Bishop (13.00), over Benarty to Blairadam (15.21), Dumglow (16.50), Rumbling Bridge (19.00). Quite a hike!

Mayday on the Cleish Hills

Alone,
With the spiralled song of the lark;
Alone,
With the winds of the wild, grey hill –
When the day is fired in the kiln of night
And eyes alive with the sights of the height.
Alone,
With the peewit's skeery screaming,
Alone,
Where the heart of the land lies still –
When the air grows cool and the heart glows bright,
When the step goes tired and the thoughts go light.

Hamish Brown

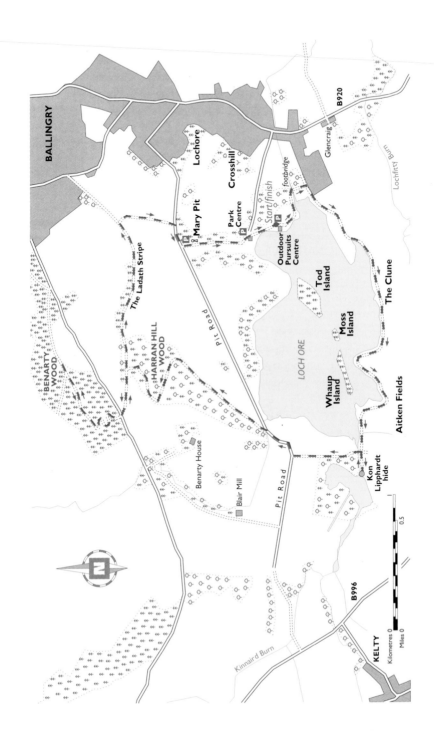

LOCH ORE CIRCUIT

Loch Ore has been recreated out of a landscape of mining dereliction and is now part of a popular Country Park offering sailing, windsurfing, canoeing, golf, riding, an adventure playground, bird-watching and various possibilities for walks, of which this is the longest and most rewarding, and one which would be excellent as a crisp winter outing or on a floral spring ramble. Few people will be encountered on this surprisingly rural circuit.

Loch Ore lies surrounded by the one-time mining towns of Ballingry, Lochgelly, Cowdenbeath and Kelty, and the only sign of its past is the monumental Mary Pit winding tower which stands in the park. The bings have gone, and all is green or forested, with the park covering 460 hectares, of which 260 are Loch Ore itself – the re-emergence of a loch drained in the 18th century. The M90 (Junctions 3 or 5) offers quick and easy access to the park, which is well signposted for motorists.

The walk begins by wandering round the quiet southern shore of Loch Ore. In winter, swans, ducks and gulls congregate on its waters. Ornithological sightings are on an update board in the Park Centre for the enthusiast and there is a hide at the western corner. Whaup Island is Scots for Curlew Island and

INFORMATION

Distance: 8km (5 miles).

Start and finish: Carpark, Lochore Meadows Country Park.

Terrain: Good paths, grassy fields and woodland tracks. No special footwear required.

Toilets: In the Park Centre.

Refreshments: Cafeteria in the Park Centre. (Also information desk.)

Opening hours: *Lochore Meadows Country Park (and Centre):* All year (buildings 0900–1700); *Cafeteria:* 1100–1615 winter weekdays (Thursday–Sunday, closed, 1330–1400), 1100–1730 and 1800–1930 every day during summer.

Winter Day on Benarty with Loch Ore below.

they are not uncommon. Swans will be seen nesting and herons stalk the reeds where great crested grebes nest. The new plantings attract many migrants including willow and sedge warblers.

The walking path actually starts from the carpark behind the Outdoor Pursuits Centre (signposted) and wiggles round to the outflow of the loch (R. Ore), which is crossed by a footbridge. At a T-junction turn right and the loch is soon rejoined. The volcanic ridges of The Clune have prehistoric hut circles and other evidence of ancient use. The path varies in character but, surprisingly, the second half is over rich, green cow-pastures, with occasional tree belts (the Aitken Fields). Something like one million trees have been planted in the park, mostly alder, pine and birch.

The prehistoric fort site on the knoll of Dunmore above Lochore.

The west end of the loch is a nature reserve, and is clearly signed. The path leads to a 'crossroads'; turn right across a footbridge, the water this time flowing into the loch. By going straight ahead first, for a few minutes, you can reach the Kon Lipphardt hide, named after a local mason involved in the building of the hide, who sadly died just before it was opened. The path, once over the bridge, runs straight along to meet the Pit Road, a tarred road, where you turn right. After an S-bend, take the first gate on the left, signposted for Harran Hill Wood. (Going straight on completes a shorter route round the loch.)

The track climbs gently then suddenly rears and twists up steeply on to and over Harran Hill,

Lochore Meadows Park with the visitor centre and boats for hire.

descending slightly and walking to a gate on to a larger track, called the Avenue. The way back to complete the round is to the right, but it is worth going left first to Benarty Wood to relish the big view that opens up. This is a steep haul but well pathed or with steps. The track goes up to the tarred Hill Road (Ballingry–M90) and opposite rise the zigzagging steps. The gradient eases and the path traverses along to join a forestry track, which is as far as you need go for the view. By turning left on the forest track, you can climb higher up and expand the view, but a return still has to be made down to Harran Hill Wood. Turning right on the forest track simply leads downhill to the Hill Road again.

Walking down the Avenue the verges are rich with wood anemones in spring (the woods are massed with wild hyacinth). The size of the Mary Pit winding tower is noticeable from up here. Leave the Avenue by a gate, right, into the field. Start down this and swing right to aim for the winding gear. A gate/stile leads back on to the Pit Road. Turn right then left at the far edge of the carpark to walk past the winding gear and an old shunting engine (pug) to regain the start.

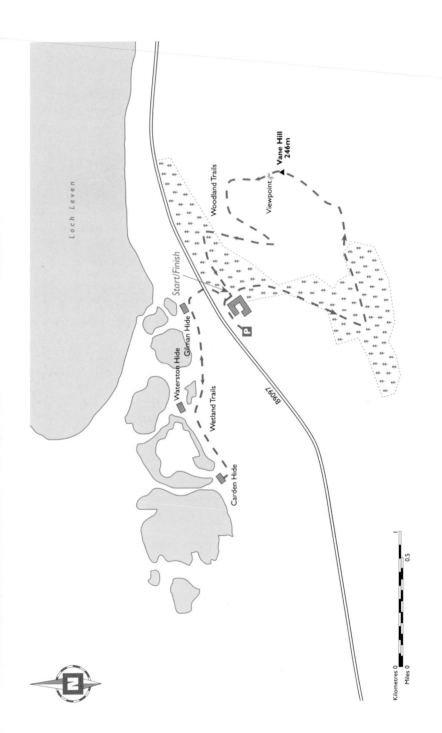

Loch Leven

Woodland Trails

Vane Hill 246m

Viewpoint

Start/Finish

Gilman Hide

Waterston Hide

Wetland Trails

P

Carden Hide

B9097

Kilometres 0
Miles 0
0.5

VANE FARM

Vane Farm was bought by the RSPB in 1967 to be an educational nature reserve and visitor centre, and it is a lively, interesting place with all the facilities for studying birds, a good shop, coffee shop and helpful staff. There are three hides near the loch, reached by an underpass below the road, which should not be missed. This is a good place to explore in early winter (October–December) when the Icelandic geese are visiting, but the varied habitats make it a pleasant spot during any season, and the nature trail gives superb views. As it is a reserve, walkers are asked to keep to the trail. Keen birdwatchers can obtain up-to-date information in the centre.

Trails are well signposted and maintained. The top of Vane Hill is 246m so the view is quite special. Loch Leven covers 14 square km and is backed by the Lomond Hills. The loch has romantic and historical associations: Mary, Queen of Scots, was imprisoned on Castle Island (to which there are summer boat trips from Kinross). You may also see more than birds circling in the thermals over these Loch Leven hills, as Portmoak (Scotlandwell) is where the Scottish National Gliding Centre is based. Scotlandwell also has a very ancient canopied well and is only a few minutes' drive from Vane Farm.

INFORMATION

Distance: 1½km (1 mile). But allow 1 hour at least.

Start and finish: At the Centre, where there is a carpark. Leaving the M90 at Junction 5, Vane Farm is well signposted and lies on the south side of Loch Leven (and is actually in Tayside, not Fife).

Terrain: Steep trail; stout footwear advised.

Toilets: At the Centre.

Refreshments: At Vane Farm during opening hours.

Opening hours: The Centre is open: January–March, daily, 1000–1600; April–December, daily, 1000–1700. Entry charge for non-RSPB members **(dogs are not allowed on the reserve).** *Loch Leven Castle:* April–September, weekdays, 0930–1830; Sunday, 1400–1830.

Looking down to Vane Farm and its wetlands with the Ochils in the distance.

Down to Vane Farm and
to the far Ochils from
Vane Hill.

The trail angles up and across by the regenerating woodlands, mostly birch and rowan, which have sprung up in the last 20 years after sheep were removed from the reserve. Willow warblers, tree pipits, redpoll and woodpeckers nest in the woods. After you leave the woods, the path climbs steeply up through bracken, heather and blaeberry to the viewpoint of Vane Hill. Meadow pipits abound and many – poor things – are fostering cuckoos in their nests. Wheatears, whinchats, grouse and curlews may also be seen. (The higher crags on Benarty have nesting fulmars.) Some of the nest boxes are occupied by pipistrelle bats.

From the viewpoint, Kinross can be seen at the far-left corner of the loch. Straight ahead, the Highland hills are glimpsed (often snow-touched well into summer). The Lomonds are bold above Kinnesswood on the right of the loch and, in the distance, right of the cut, are the twin tops of Largo Law. The Bass Rock and North Berwick Law sit high beyond the estuary of the Forth. Beyond Benarty lie the Cleish Hills then a view west to the distant hills of Loch Lomond and the nearer, bold Ochil scarp.

The island nearest to the centre of the loch is St Serfs, on which you can see the ruins of an old priory. Ducks nest on the island. The loch level was lowered by 1.4m in the 1830s when the long, straight cut was made to control the Leven's flow for the many mills downstream. The lagoons or scrapes are man-made to attract wading species and winter visitors (geese by the thousand, whooper swans by the hundred). It is worth walking out on The Wetland Trail to visit the hides, but, beware, people visiting Vane Farm simply to walk are apt to leave as nascent ornithologists!

Sunset on Vane Farm.

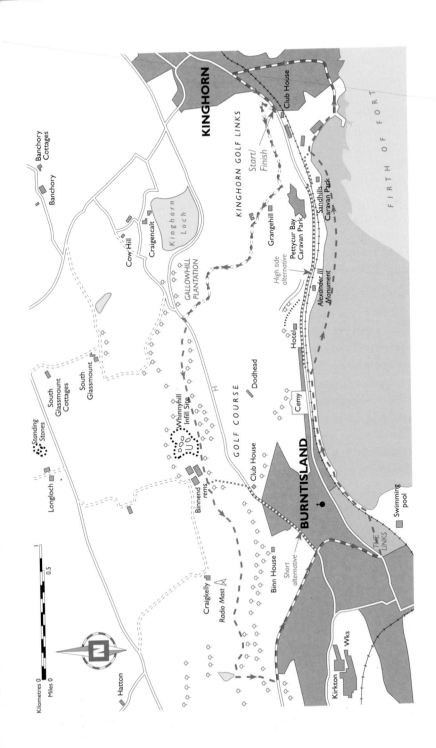

THE BINN

The start is at the Kinghorn Golf Club building. If coming by bus ask to be dropped off there, otherwise walk west from the centre of Kinghorn (war memorial by Alexander Carrick) and up the steps or the drive facing the 1895 clubhouse. The track swings left at another clubhouse to rise steadily across the golf course. (Does hitting a pedestrian count as a penalty, I wonder?)

At the end of the golf course (Pettycur Bay Caravan Park on the left braes) the track swings inland, passing Grangehill House (left) and a farm beyond (right). Note the interesting 'window' on the south gable of the last building. All along this walk there are many rabbits and, hereabouts, black ones are not uncommon. Looking ahead, the Binn mast appears deceptively near.

As the track turns more inland, the seaward setting is forgotten awhile; cows browse on the rolling slopes and skylarks sing overhead. At another bend, though, the sea (Largo Bay) appears again, off right, with Kinghorn Loch, in its secretive hollow, below. The red slopes are waste alumina from the Burntisland aluminium works.

When the old Kinghorn–Burntisland road is reached turn left and, after 30m, cross carefully to a gate/stile beside the fenced-off area. This long, gradually ascending track is the old road to the Binn village, at one time a thriving community of 1000 people, owing its existence to the shale oil industry, now completely gone. (Under here and under the Forth and West Lothian lies a warren of old shale oil workings.) Halfway along the ascent, the water on the right goes across and down to a settling

INFORMATION

Distance: 10km (6 miles).

Start and finish: On the street by Kinghorn Golf Club.

Terrain: Tracks, paths, steep, grassy areas and sandy beach. Strong footwear advisable.

Toilets: Kinghorn and Burntisland.

Refreshments: Kinghorn and Burntisland, and Kingswood Hotel, between the two towns.

Opening hours: *Burntisland Museum:* during library hours. *Parish church:* by appointment (details from Kirkcaldy Tourist Office, tel: 01592 203154).

The Binn seen over the Alcan works, Burntisland.

pool. Look along the line of the ditch and, on the golf course, just over the road, you'll see a shelter made from half a boat. Aluminium and related products go into things as diverse as cosmetics, paint thickener and oven foil. The works will be seen from the top of the Binn.

A gate indicates the end of the long track up, and the ruins of the village lie on the right. Left, a gap in the wall and steps indicate a path down to the road if a shorter route is required, but two stiles and a couple of fields now lead to the top of the Binn, a viewpoint quite marvellously fine for its modest 193m (622 ft). The cliffs fall steeply to Burntisland below and, being crumbly basalt, are propped up with concrete buttresses, avalanche barriers and a chain-mail of anti-submarine netting. There's a view indicator on the summit.

Behind the big mast is Benarty, right of it are the cones of the Lomonds and left, behind the billowing clouds created by Mossmorran, are the lumps of the Cleish Hills – all walks described elsewhere. Ben Chonzie, 40 miles off, is a Highland Munro, and in the west the hills above Glasgow are clear.

Before leaving the summit, gauge the state of the tide. If it is up to the curve of the railway with no sand showing then the return to Kinghorn should be made along the main road. This passes the monument to Alexander III who was killed when he was thrown off his horse and fell over the cliffs in 1286 while he was hurrying home – a sad event in Scottish history.

The route down keeps along the cliff edge, accompanied by wind-contorted thorns, and over a grassy rise as high as the Binn itself. In the hollow beyond, bear right and then down towards a lochan, picking up an old 'green road' round and down to the coot-noisy waters.

Go through a gate then turn left at a footpath sign to another gate and the path leading across the field to a wood beyond. The path down through the wood to the road indicates old quarry workings here. Nature heals surprisingly quickly.

Turn left and walk down to the roundabout and on to Cromwell Road (Cromwell won a vital battle a few miles west in 1651 and garrisoned Burntisland). This leads down to the Links (where a summer fair is held) with a cherub fountain at the town end. The Old Port marks a town gate from the past; note its triple, mottoed sundials and other adornments. There are several cafés and pubs on Burntisland's High Street – which will probably be welcomed.

On Kinghorn's braes in winter.

During library hours do look at the museum upstairs, laid out as a Victorian funfair. Burntisland's parish church is the earliest surviving post-reformation church in Scotland and has many interesting features. Also of historical note is that in 1601 the Church Assembly met here and decided that there should be a new translation of the Bible – this became known as the King James version.

Once refreshed and Burntisland explored, head over the links to one of the tunnels under the railway to gain the promenade. This ends at a house with a railway underpass beyond. If the tide is out, miles of sand lead back to Kinghorn, if the tide is in, pass under the bridge and take the coast road past the monument to Alexander III. (At the Sandhills Caravan Park a footpath leads back down to Pettycur Bay.) Pettycur was once the main crossing point for travellers heading north. The road up from the harbour leads back to the start.

A shorter walk – though still enjoying the fine views from the top of the Binn – is a round walk from Burntisland: going out the old Kinghorn road (Kirkcaldy Road) to the golf club and the path opposite up to the Binn village and back as already described.

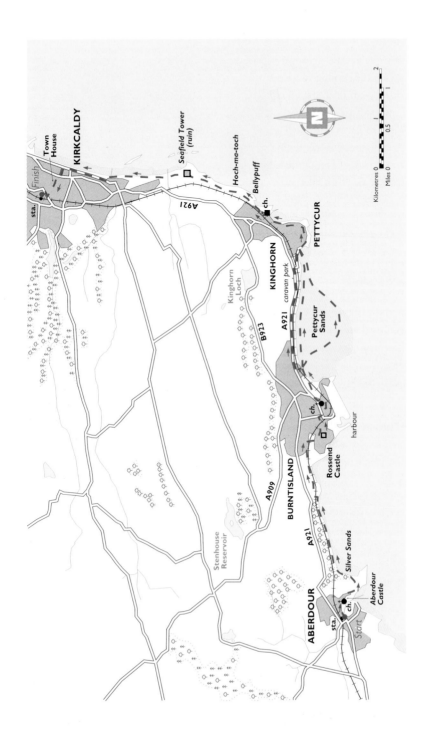

THE CASTLES COAST

This is a varied and quite demanding walk so, while looking at sites and other places of interest, don't linger too long. The Kirkcaldy Museum and Art Gallery is popular and has much to see, so be there at least an hour before closing time. The station is immediately behind it, so returning to Aberdour is easy. If time becomes short you can finish at Kinghorn station. Aberdour won the Best-kept Station in Britain Award in 1990 and is a floral delight. Beside its entrance is the drive leading to Aberdour Castle (Historic Scotland), which is the first of several castles of note on the walk. The Earls of Morton owned Aberdour Castle – one of them had his head cut off by the Maiden (guillotine) and another imprisoned Mary, Queen of Scots, in Loch Leven Castle. Only part of the building survives but there are several fine sundials and a notable beehive doocot.

From the castle go through to see the old church and its graveyard before heading on down to the Silver Sands and the well-made path along the coast to Burntisland; a woody walk by sea and railway until the huge complex of the aluminium works is reached. Turn right halfway along its perimeter and up the hill to reach Rossend Castle, a derelict tower that has been restored and stands above Burntisland Harbour. It has its Mary, Queen of Scots, association too. The hapless French courtier Pierre de Chastelard was found under the Queen's bed (for a second time) and lost his head rather literally thereafter.

Until the Forth Bridge was built, a train ferry crossed to Burntisland, and both Cromwell and

INFORMATION

Distance: 16km (10 miles).

Start and finish: Aberdour to Kirkcaldy. Regular train and bus services link Aberdour and Kirkcaldy. Parking at stations. Public transport information line, tel: 01592 416060

Terrain: Mostly on paths but rough in places so stout footwear needed.

Toilets: In each of the towns passed and some of the sites mentioned, and also the museum or station at the end.

Refreshments: At Aberdour Castle, in the towns on the way (cafés and bar meals) and the museum at the end.

Opening hours: *Aberdour Castle:* weekdays, 0930–1800. Sunday, 1400–1800. *Burntisland Museum:* Library hours. *Burntisland Church:* details from Tourist Office, tel: 01592 203154. *Kirkcaldy Museum and Art Gallery:* weekdays, 1030–1700; Sunday, 1400–1700.

Burntisland Harbour and Rossend Castle.

the Romans thought it a useful port. Work down to the High Street and along to the library building which houses a museum based on a Victorian fair (the fair continues to this day every summer). Round the corner, if you walk up and turn left, you find the gates of the historic church. Visits can be arranged through Kirkcaldy Tourist Office. There is no stained glass so the interior is light and airy, box pews survive and balconies bear heraldic and trade symbols while an outside stair allowed sailors to enter quietly and discreetly.

There are several cafés at the east end of the High Street. The Links are then crossed to pass under the railway and back to the shore. See if you can spot the crocodiles on the cherub drinking fountain.

If the tide is out, you can walk along the rim of the huge Pettycur sands to the little harbour of the same name. If the tide is too far in then you will be forced to follow the road, which passes the monument to Alexander III, last of the Celtic kings.

For centuries Pettycur was the main Forth estuary crossing for those heading up the east coast or to St Andrews, and this explains the 'Pettycur' on so many Fife milestones. Walk up Pettycur Road and turn off, right, at Doodells Lane, which will lead you round to Kinghorn Bay where the unostentatious 1774 church is sited. The way on is blocked (there was once a thriving shipyard) so head up by the church, under the railway and then right at a play area on to a good path for Kirkcaldy.

If wanting refreshments go right up this road and left on to the High Street. (The continuation inland has some attractive 18th-century houses with marriage lintels, forestairs and pantiles.)

Hoch-ma-toch and Bellypuff are names attached to the first bay but note the rocks: lavas flowing down over limestone and shale levels as if they had only cooled yesterday. Wend along – and up – and cross a wall, as grassy slopes lead down to the stark, red ruin of Seafield Tower. There's a shoreline well, and seals often haul themselves on to the rocks just offshore and 'sing'.

Beyond the castle the path becomes a track and there is redevelopment taking place on what was once the site of the huge Seafield Colliery. There were bleach works, a rope factory, potteries, a sweet factory and other industries, too, all now swept away. At the grassy area you can exit to the main road and walk round to the esplanade, though at low tide you can walk the sands along to the bulge in the sea wall. You will see why Kirkcaldy (pronounced Cur-coddy) is called the 'lang toon'.

Seafield Tower.

Walk along the promenade as far as the traffic lights then turn inland and right along the High Street. When it becomes pedestrianised swing left and walk straight on, passing the Tourist Information Office then the fine Town House and Bus Station (right) and Sheriff Court (left) and crossing at a T-junction to reach what are war memorial gardens. The Museum and Art Gallery building lies at the top end of the gardens. Over on the right is the Adam Smith Centre with halls and a café.

The museum is a busy place and has several exhibitions each year besides its prize-winning local room, some famous paintings (Peploe, McTaggart, Hornel, among others), a shop and a café where you sit amidst showcases of glittering Wemyss Ware.

Adam Smith Centre, Kirkcaldy.

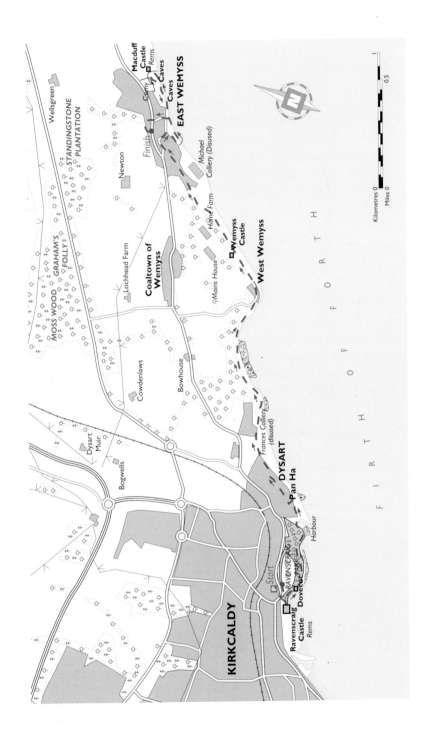

KIRKCALDY

Ravenscraig
Castle
Rems

Dovecot
RAVENSCRAIG
Harbour

P Start

DYSART
Pan Ha

Frances Colliery
(disused)

WEST WEMYSS

Wemyss
Castle

Mains House

Home Farm

Michael
Colliery (Disused)

Coaltown of
Wemyss

EAST WEMYSS

Caves
Cany
Rems
Macduff
Castle

Finish

Newton

Wellsgreen

STANDINGSTONE
PLANTATION

MOSS WOOD
GRAHAM'S
FOLLY

Lochhead Farm

Cowdenlaws

Bowhouse

Dysart
Muir

Bogwells

FIRTH OF FORTH

Kilometres 0
Miles 0

0.5

DYSART AND V

This is another section of Fife's Castl Coast and, like Walk 8, gives plenty of things to see, as well as letting you enjoy the walk along by the Forth Estuary with its seascapes and wildlife. It would have merited the name the 'Coal Coast' 50 years ago.

Before leaving Ravenscraig Park follow the signs to Ravenscraig Castle. The setting is splendid and Sir Walter Scott used it in his *Lay of the Last Minstrel*. It is important, architecturally, being the first built specifically to cope with the then new invention of cannons. This accounts for the thick walls (14 feet in places) and the sloping roof on the tower (so cannon balls skidded off). Ironically, James II, who began the work of building the castle, was himself killed by an exploding cannon when besieging Roxburgh Castle in the Borders.

Once out of the castle, head eastwards through the park, keeping close to the seaward side where there's a ridiculous zigzag wall, which was built – spitefully – by a one-time landowner to stop workers walking from Dysart to Kirkcaldy along the shore. (He went bankrupt over gambling debts and the next owner presented the park to Kirkcaldy!) There's a well-preserved doocot as, anciently, fresh meat was rare in winter and every laird's castle had a doocot by it.

Eventually you will find yourself looking down directly on to picturesque Dysart, an unexpected and attractive view. Carrying on, you reach a walkway and steps that lead down to the harbour. The 'tower' is actually the old St Serf's Church, not a castle, though it could double as a defence. The attractive white houses with the red pantiles line Pan Ha (*ha* = haugh = level area; now grassed). The pans were used for making salt – for which Dysart was once famous, as it had the other vital ingredient to hand: coal.

Walk along Pan Ha and then up the steep Hie Gate to reach the town centre. The John McDouall Stuart

Distance: 9km (5½ miles).

Start and finish: Ravenscraig Park, Pathhead, Kirkcaldy (large carpark) lies east of the main town and is the best start. Ensure the Kirkcaldy bus from the finish goes by Pathhead. For transport details call 01592 416060/642394.

Terrain: Mostly on paths or streets. Stout shoes adequate.

Toilets: In the towns, but none at Ravenscraig Castle.

Refreshments: Cafés in towns and the Belvedere Hotel, East Wemyss, is strategically placed for walkers following this route.

Opening hours: John McDouall Stuart Museum: Dysart, June–August, 1400–1700. Wemyss Environmental Centre: East Wemyss Primary School basement is open during school hours, 0900–1630.

Ravenscraig Castle.

Museum commemorates Dysart's most famous son. Born in 1815, he spent much of his life exploring Australia and was the first man to cross that great country from south to north. The Cross has a Victorian lamp marking it and a very photogenic tolbooth, with a 1576 date over the outside stair. Before you walk on, turn left up Cross Street to see the stylish 16th-/17th-century houses opposite the end of the street. Heading east from the cross the road climbs somewhat and when it swings left (past council houses) bear off right to follow a path that passes the winding tower of the closed Frances Colliery – a stark reminder of an industry that once dominated this coast. Leven power station was built to use up the slurry (successfully), and Fife Enterprise has landscaped the coastal stretches ahead. Woody slopes lead to a prow with a view to West Wemyss. Drop steeply down and along to this attractive, quiet corner.

Among the restored buildings round the old harbour (now largely infilled) the one-time Miners Institute has been turned into the Belvedere Hotel (which has good bar meals and old pictures of local interest). All signs of the mines have gone and many buildings in West Wemyss are empty, though you don't notice till you look carefully. Halfway along the Main Street there is a tall, slender tolbooth tower, with a pend (passage) through to the back. Tolbooths were customs points, gaols and town houses, and, like doocots, Fife has the lion's share of surviving examples. The swan symbol on the tolbooth is the heraldic device of the Wemyss family, whose castle dominates the nearby shore.

The town ends at St Adrian's Church, whose graveyard has some stones of interest. One is made of coal and one has the date of death as April 31! Walking on eastwards, pass under the grim-looking Wemyss Castle (Wemyss comes from the Gaelic word for 'cave') and the walk will end at the famous caves which lie beyond East Wemyss. After a straight section of path, instead of turning right at the coastal path sign, take the wooded left fork to reach East Wemyss at the architecturally interesting rows of miners' houses. Below, now planted with trees, was the site of the Michael pit, which suffered a disastrous fire in 1967. Steps lead down then cross the grassy area to Approach Row, and from its far end follow the road down to the shore again, near the old parish church war memorial with the miniature soldier. The graveyard has some good trade-symbol stones.

Heading along the shore, a set of display boards describes the mainly unsafe, but fascinating, Wemyss Caves. Much has been lost due to roofs collapsing (the danger warnings are to be taken seriously), but it is safe to enter the Doo Cave with its rows of rock-carved nesting boxes. Below Macduff Castle, the last on the Castles Coast, are more caves, and Jonathan's Cave contains perhaps the best carvings. Look for the oldest representation of a ship known in Scotland (back, east wall). The cave is named after a one-time nailmaker who worked in it.

Doo Cave, Wemyss Caves.

From the caves, walk back westwards to turn up East Brae towards the coast road (and the bus back to Pathhead). On the way, it is worth diverting into the large cemetery lying above the caves and Macduff Castle. Stones tell of disasters in the pits, at sea, in the wars and on the mountains. There is a memorial designed by Charles Rennie Macintosh, and a cherubic figure marks the grave of Michael Brown, 'done to death' in a sensational 1909 murder. Booklets on the caves and the Michael Brown saga are on sale at local museums or the Book House, Tolbooth Street, off Kirkcaldy High Street.

Coast buses will take you back towards Kirkcaldy, but ensure you get off at Pathhead, as most buses do not actually pass Ravenscraig Castle.

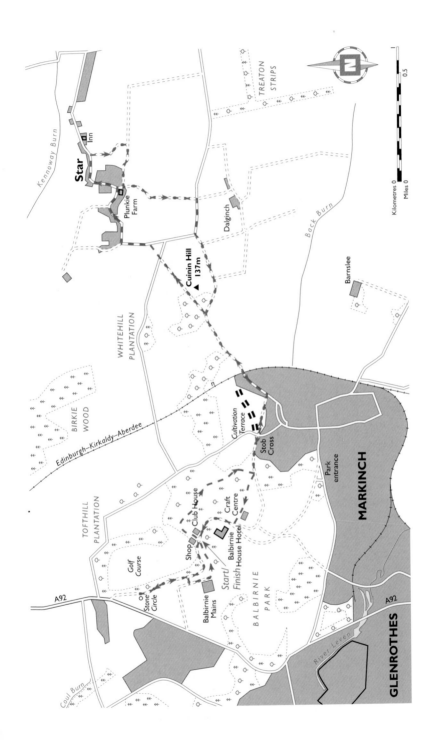

Star

Kennoway Burn

Inn

Plunkie Farm

Dalginch

▲ Cuinin Hill 137m

TREATON STRIPS

Back Burn

Barnslee

WHITEHILL PLANTATION

BIRKIE WOOD

Edinburgh–Kirkaldy–Aberdee

Cultivation Terrace

Stob Cross

Park entrance

MARKINCH

TOFTHILL PLANTATION

Club House

Craft Centre

Shop

Golf Course

Balbirnie House Hotel

Start/ Finish

Stone Circle

Balbirnie Mains

BALBIRNIE PARK

A92

A92

River Leven

GLENROTHES

Coul Burn

Kilometres 0

Miles 0

0.5

BALBIRNIE AND STAR

Balbirnie Park has many fine specimen and exotic trees (60 different species) and a wealth of rhododendrons collected from the Himalayas, so it is worth visiting in early summer.

Markinch lies east of Glenrothes and is best reached from the A92. Turn off the road facing the Tullis Russell Paper Mill entrance, which is signposted for Balbirnie Park, and the entrance is on the left as Markinch is reached. Follow the drive past the mansion (Balbirnie House Hotel) and use the extensive woodland carpark beyond. The prestigious hotel is open to non-residents, and a meal, tea or refreshments there may be welcome at the end of your explorations.

The 416-acre park is the old Balfour family estate, which prospered in the 18th century with profits from improved agriculture and coal mining. Head off past the golf shop, clubhouse and carpark to be welcomed into the woodlands by golden yews.

Stay on the right bank of the burn after an open area; the path climbs and gives views across to some soaring Wellingtonias. Ignore paths heading off right or left. Where the woods end, an area once-derelict from mining has been reclaimed. On entering the old woods again, turn left at the first opportunity and, bearing left, walk out to the East Lodge. Turn

INFORMATION

Distance: 8km (5 miles).

Start and finish: Balbirnie Park, Markinch.

Terrain: Parkland and rural landscape, mostly on paths or tracks, but waterproof footwear advisable.

Toilets: Balbirnie House Hotel carpark, Plough Inn at Star.

Refreshments: Balbirnie House Hotel; Plough Inn, Star.

Star.

right, crossing to the narrow pavement.

Look back to see the ancient Stob Cross up on a bank, a rare example of a sanctuary cross connected with the parish church, the spire of which soon appears ahead. The site had an early church built by St Drostan, nephew of St Columba, and Markinch may have been the capital of Fife when it was one of the seven kingdoms of Pictland. The tower of the church is 13th century.

Not far past the Markinch town sign, the route goes through a gap in the wall and along past a play area to reach a street where you turn left (50m further on and on the right there is a Victorian letterbox). Walk along and under an echoing railway bridge (the Edinburgh–Kirkcaldy–Aberdeen line) for the signposted path to Star. Left is a Victorian graveyard with a splendid array of yew trees. The Celtic cross for the Balfours of Balbirnie is the most interesting feature.

The Balfour cross in the North Hall graveyard.

The footpath for Star rises up behind the cottages, a steady pull. There's a long straight and just 40m after the path bends, leave it to take the stone steps up left for a path through the woods of Cuinin Hill. At the top of the wood a kissing gate gives you access so that you can walk along the edge of a field. The view expands, with Largo Law prominent and the village of Star straight ahead.

Cross the first road and then a path opposite cuts the corner of the field to reach the road into Star. This is a charming village of cottages that has an almost unique rural atmosphere. Follow the twisting road through the village. Make a note of Plunkie Farm (right, red pantiles) where there's a footpath sign for Dalginch. You will follow this after visiting the far end of the village, joining it by another footpath starting 100m beyond the post office (signed for Plunkie Farm/Treaton).

Towards the end of the village – just where walkers will welcome it – is the Plough Inn which has snacks, bar meals and a restaurant (open midday–1430 and 1800–midnight). Turn back and go up the Treaton/ Plunkie Farm track; it wiggles past some houses and then swings right to run along to Plunkie Farm. Turn left and follow the farm track that leads up to a minor road. Turn right and, at the first sharp bend, break off left on a farm track which you follow along (ignore the track going left after 50m) to the tidy-looking North Lodge. After that it becomes a footpath and joins the outward route from Markinch.

Return to Balbirnie Park by the East Lodge, but walk straight on from it, ignoring other paths and the outward route. After a 'crossroads' take a path that angles up a grassy bank, right. It joins one contouring Fir Hill where there is a magnificent *Pinus nigra* with a seat around its base and backed by a row of monkey-puzzle trees. Walk left above the old glasshouses (behind the holly hedge) to come out at the Balbirnie Craft Centre, created in the former stable block.

Some workshops are open to visitors and there's a fine art gallery in the corner of the yard. Coming out from the yard, turn right then the first opening on the left (where there is variegated holly) will lead back to Balbirnie House.

Before stopping for the day, there is one more short walk not to be missed. Continue along past the carpark to a gate (at a bridge) and follow on up the Balbirnie Burn (in early spring there is a mass of snowdrops under the trees), cross a gated transverse track and, at a grassy area, there is a prehistoric site of some interest. An interpretative board explains the attractive site. Walk back

Snowdrops in Balbirnie Park.

down to the two gates, turn right and cross the bridge to walk down the other bank of the burn that leads back to the start.

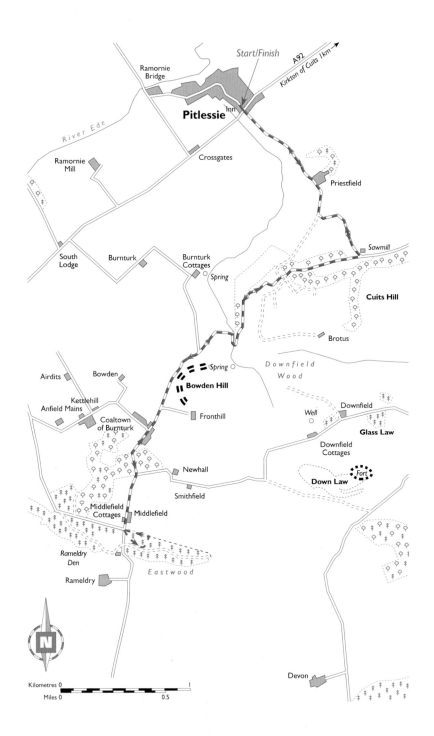

ABOVE THE HOWE OF FIFE

INFORMATION

Distance: 9km (5½ miles).

Start and finish: Pitlessie, on the A92 (6km west of Cupar). Park in village.

Terrain: Tracks and tarred road. No special footwear required.

Toilets: Pitlessie only.

Refreshments: Inn at Pitlessie or Cupar (6km).

This is a gentle walk that gives sweeping views of Fife's heartland. It is something of a mystery tour, too, and is a good walk for a winter's morning or an evening when the light can flood along the Howe of Fife. Pitlessie is a charming village forever associated with the painter Sir David Wilkie who grew up there and, in 1804, painted *Pitlessie Fair* (National Gallery of Scotland). The now-converted maltings point to the village's past history. Pitlessie is a 'best-kept village' contender and typical of the Howe of Fife, as the great level sweep by the River Eden is called. With the river (and old Rossie Loch), drained in the 1740s and 1800s, the land is some of the richest in Scotland.

Cross the busy A92 from the historic Pitlessie Arms to head up the good farm track to Priestfield, passing right of the farm to curl up on to the 'balcony road', as I've heard it called. Join it through old lime workings where there's a huge concrete limekiln with a plaque 'Rebuilt 1937'. On the left is a sawmill, and in it you can see a surviving outcrop of limestone. Quarrying took place all along this 'balcony' of Cults Hill for centuries but has largely died out and now little is seen, so vigorous has been the regeneration of trees. Beside the crossroads, 1km east, however, there is still a busy brick and lime

The old coaching inn at Pitlessie.

Unusual warning sign at Burnturk.

works, and more or less opposite where you come out on to the tarred road is the entrance to one quarry still in use. This whole upland area is called the Riggin o' Fife.

Turn right along the road. After about 500m the trees end and there are superb views over the Howe to the Lomonds and the northern rim of Fife. There is a hairpin bend and then the road circles Bowden Hill (a prehistoric fort site) to reach the scattered hamlet of Burnturk. Beside its town sign is a triangle warning sign with a frog on it! (It is only displayed 'in season'.) The fuller, older name was Coaltown of Burnturk. Coal often lay conveniently near limestone measures.

Keep on ahead, passing a wood then open country, to reach a crossroads (signs for Kingskettle, Star, Burnturk). Turn left on the old coffin road to Devon and just past the third small parking space there is a path leading into Eastwood trees. I won't divulge why – or what you will find in there. The footpath makes a circuit in the wood to come out further along the track and then you walk back to Burnturk and the 'Balcony' road. As this curiosity is not publicised I trust readers/walkers will leave others to be intrigued, and surprised, in turn.

The walk back is along the outward route (there are no safe, practical alternatives) but you gain a whole new perspective and the views vary from hour to hour or minute to minute. Visitors, when driving, really should take time to explore some of these minor roads in Fife, both for the scenery and to see some of the sites and sights.

Over the Howe of Fife.

Back in Pitlessie, there is a post office/village shop and an inn to offer refreshments, and then a short historical addition can be made by car. Turn east on to the A92 (Cupar direction) and, after 1km, turn off right at a sign for Kirkton of Cults. Cults church was rededicated in the 12th century. The present church, however, with its ornate birdcage bellcote (which is made of old gravestones) and outside stair, is late-18th century. Inside, it still has box pews and looks much as it must have a century ago – including long ladles for taking up the collection. David Wilkie's father was the minister, and there are many other family plaques – his relatives going to India and Australia, all very much part of that period's social history. In the cemetery there are several Cochrane family stones, a family related to the Earls of Dundonald, noted at Culross. One stone commemorates a Cochrane couple happily married for 70 years!

About 1 km to the east there is a tomb (not accessible) on a hill called Lady Mary's Wood where the Victorian builder of Crawford Priory mansion is buried. She was noted for running about the woods with the deer, starkers. Strictly speaking she rebuilt a reasonable lodge, turning it into an architectural hotchpotch. The gutted shell still stands. Lastly, before leaving, note the fine early-18th-century doocot and the huge size of the one-time manse. In centuries past, manses were often used like inns; no doubt pigeon pie appeared on the menu. The tiny building at the entrance is the session-house. Both the walk, and wandering around here, will have given you some idea of past life in rural Fife. The folk museum at Ceres (Walk 15) would be a good follow-up day and the Hill of Tarvit (National Trust for Scotland) is one mansion that can be visited.

Harvesting in the Howe of Fife above Pitlessie.

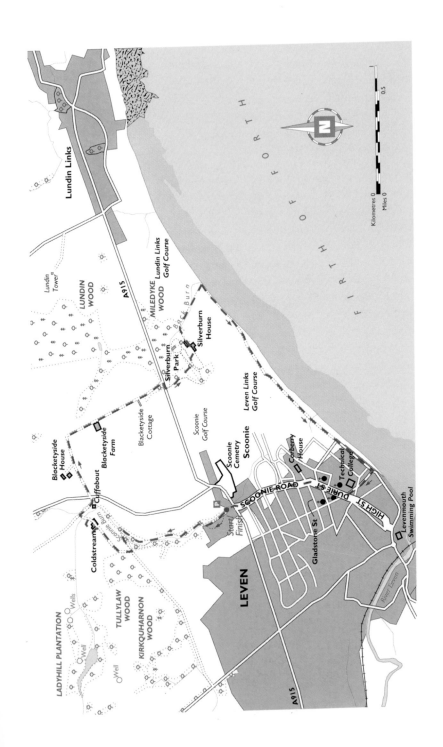

LETHAM GLEN AND SILVERBURN

This is an astonishingly varied walk and offers much for children, as two places have animals and birds on show. The complete circuit, however, should not be undertaken by small children, who, instead, would enjoy shorter variants or separate visits to Letham Glen and Silverburn. Winter can often be a good season for taking this walk at a brisk pace, and the glen and woods are colourful in autumn.

Letham Glen gives a very pleasant garden and woodland walk. Head up the main drive as far as the Pets' Corner (aviaries, animal enclosures). Cross the bridge beside the doocot/rabbit warren and wend on upstream to the end of the glen. Cross the burn and take the highest path to angle up to a park entrance. (The option of a short circuit can be made by walking down the glen again.)

Turn right on a farm track. When it swings right there is a grand view of Largo Law. Circle round the farm (Coldstream) and out by the track beyond. A small burn is crossed and a bigger burn, the Letham (which becomes the Scoonie lower down), is followed downstream to a house called Cuffabout. Cross the bridge to reach the Leven–Cupar road and go straight across on the Blacketyside House private road. Past this attractive house the

INFORMATION

Distance: 6km (4 miles).

Start and finish: Carpark at Letham.

Terrain: Good paths, tracks and roads. No special footwear required.

Toilets: Letham Glen, Silverburn and Leven Promenade.

Refreshments: Leven; many facilities.

Opening hours: *Silverburn Mini Farm:* Monday–Friday, 0830–1630, weekend afternoons. *Arts and Craft Centre:* weekend afternoons; Wednesday. afternoon (July, August). *Letham Glen:* daylight hours.

The doocot/rabbit house in Letham Glen.

Blacketyside cottages.

road turns sharp right to pass a big farm then some fruit fields and attractive cottages, one of which is an Embroidery Workshop and Gallery. (Open: Sunday, 1300–1700; Tuesday-Saturday, 1000–1700. Tel: 01333 423985.)

Cross the busy A915 (great care needed) to reach Silverburn Park, which has gardens, trails, specimen trees, a mini-farm and an arts and crafts centre. Take the gate opposite the entrance (rather than the drive) and follow the footpath straight on to a T-junction with a bank beyond and views of the sea. Turn left, along a track that merges with another, and, not far short of the open area by Silverburn House (the arts and crafts centre), go through a sturdy gate on the right. This path descends through the attractive walled garden and a gate will allow an escape on to an avenue with the mini-farm opposite.

The farm has a large collection of farm animals, various species of fowl (e.g., Scots Dumpy, Indian Runner) and a display of old machinery. Once out, turn right, go past a row of red-roofed cottages and into the carpark, right. From the far corner a path leads down to the sea, following the line of the Back Burn and a wall (the Mile Dyke), which separates the Leven and Lundin Links golf courses. Once a year the clubs combine for a game, using half of each club's course and being entertained in the appropriate clubhouse at the halfway stage.

Walk westward along the edge of the golf course (or on the sands) to reach a caravan park. There is a path beside it (on the landward side) and then you are back on the Leven promenade. Leven itself is the market town for this area of Fife so is well provided with shops and refreshment choices. Note the Beehive plaque on the building at the top end of the pedestrianised High Street, and at the seaward end (near the huge chimney of the power station) is the very modern Levenmouth swimming-pool and Sports Centre; tel: 01333 429866. You could fill a day quite easily with the full walk, with its attractions, and enjoying the town.

To reach the town centre, walk along the promenade and turn right at the roundabout. Walk up past the technical college. The road swings right (by pedestrian lights), and this is the way back to the start, while left is the pedestrianised High Street, starting at the building with the Beehive plaque above.

Walking back to the start you will see a model of the beehive up by the clock, left, but they forgot to depict any bees! Dune Street, becoming Scoonie Road, has an incredible number of churches, as well as the war memorial, and opposite Gladstone Street is Carberry House where the multifaced sundial, which was once the town's mercat cross, has been erected – a very unusual type.

Silverburn – grooming time for the Shetland ponies.

Back at the roundabout outside Letham Glen, do turn up the hill and go in to explore the Scoonie graveyard. There are some notable old stones – including one with a beautifully carved ship. The best stones are at the Scoonie end of the site. Leven and Scoonie were once separate parishes and only joined up this century.

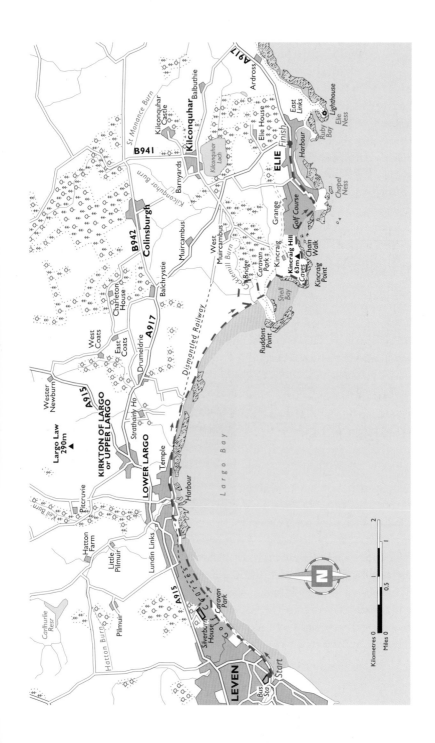

LARGO BAY

This is a delightfully varied walk for it incorporates the first of the East Neuk towns with their special atmosphere as well as allowing for remote walking on dunes – it also offers the energetic the oddity of the Chain Walk.

Wherever you park in Leven, head east along the promenade to the golf course and the Leven Beach Caravan Park. It can be passed on either side and both paths follow the perimeter of the golf course. At the first burn (bridged) you can turn inland, along the Mile Dyke, to visit Silverburn, a centre based on the old Russell estate with its flax mill, which is described in Walk 12.

Back at the Mile Dyke, you join Lundin Golf Club's course and walk along its edge with a fine view of Largo Law (Walk 14) ahead. At the far end keep to the shore, unless the tide dictates following an inland variation. Waders often congregate on the shore during migration times and winter. A path east from the golf clubhouse leads you to Lower Largo, with its rivermouth setting backed by the old railway viaduct and the Crusoe Hotel situated in the old granary building.

Halfway through the old town you'll see why the Crusoe name appears here: Lower Largo was the birthplace of Alexander Selkirk – a vivacious lad who was eventually packed off to sea, where he did very well, and actually *asked* to be marooned on Juan Fernandez Island, so bad a ship was he on (it sank not long after). Probably, he hadn't bargained for a

INFORMATION

Distance: 16km (10 miles).

Start and finish: Leven is a busy town for the central Fife coast and has a bus station facing the swimming-pool complex at the mouth of the River Leven. Parking is in a big, central carpark or on the promenade. The walk finishes at Elie, from where there is a bus back to Leven. (See note at end of walk.)

Terrain: Streets, seashore, dunes, coastal path and optional rock scrambling. Boots suggested.

Toilets: Leven promenade, Silverburn, Lower Largo, Elie.

Refreshments: Only at Lower Largo (Crusoe Hotel) and Elie.

Opening hours: *Silverburn*: Open office hours, weekdays, and afternoons at the weekend. (Walk 12 gives fuller details.)

Lundin Golf Club course.

Lower Largo.

four-year-and-four-month stint alone, but, once home, Daniel Defoe made use of his knowledge and wrote his famous book, which is based only marginally on Selkirk's experiences.

The road diverts around what was once the Cardy net factory (or use the footpath that leads straight on through to a carpark). Beyond this you are in Temple, as the east end of Lower Largo is called, possibly because of an early link with the Knights Templar. As the town ends you carry on along the empty sweep of dunes backing fair Largo Bay.

The bay ends at the Cocklemill Burn, which may be too deep to ford if the tide is in (at low/halftide you can paddle across). The burn loops through salt marshes where sea asters bloom, and there is now a strong bridge built across it as part of the official Fife Coastal Path. Skirt along the side of a camping/caravan site on the well-named Shell Bay with its big dunes.

Looking ahead, Kincraig Point looms and there are three distinct terraces of raised beaches showing what were one-time sea levels. The easy line wanders round to the point and then up these natural steps on to the top of the 63m cliffs of Kincraig Point (radio mast and wartime defences still clear). The line keeps high for a while and then drops down to a golf course. The Chain Walk alternative keeps to the shoreline and gives an energetic challenge – assuming the tide is far enough out (all the chains have been renewed recently).

The Chain Walk has carved steps and vertical or horizontal chains strung on the difficult sections of the cliffs, so is demanding (and no place for dogs), but it does allow you to see several caves, some spectacular displays of columnar basalt and other coastal features, besides being great fun in itself. If tackled, allow a full hour for this passage.

When either route chosen reaches the golf course, follow along its seaward edge (or walk the warm, gold sands) to reach a sunken track. Take this inland to reach Earlsferry, which runs on into Elie without any break so it feels like one long town. The church, with its distinctive 1726 clock tower, marks the centre of Elie (it has three clock faces only as there were no buildings to the north when the clock was added in 1900), and there are some intriguing old stones on the wall by its door. The town is clean and fresh with golden, sandy bays, a harbour, which is much used by sailing and windsurfing enthusiasts, and several excellent restaurants.

On the Chain Walk.

Note: buses for Leven stop opposite the church. Check times in advance if possible. (Kirkcaldy Bus Station, tel: 01592 642394 or St Andrews Bus Station, tel: 01334 474238, or look at the timetables displayed at Leven Bus Station before starting.)

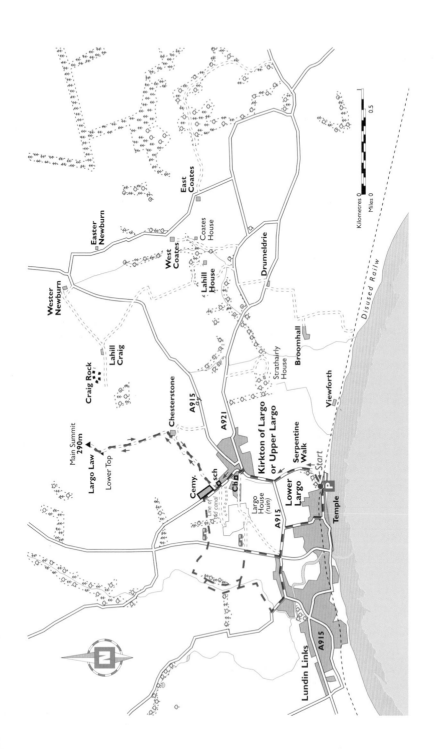

LARGO LAW

L ower Largo, on the coast, is linked with Lundin Links on the A915, so turn down for Lower Largo where indicated and head eastwards. The rivermouth area has the Crusoe Hotel (originally a granary) and the high arches of the now-defunct coastal railway.

Further along is the famous statue of Alexander Selkirk ('Robinson Crusoe') who was born here. The single street is narrow and has a couple of tight-angle bends to come right to the sea's edge. The Temple carpark is on the left and a notice-board relates some of the historical facts about the area.

Walk east for another 150m where the start of the Serpentine Walk is signposted. This crosses the old railway line (now a walkway) and continues up to woods owned by the Woodland Trust. Notice how the trees are bent, west to east, by the prevailing winds.

The path comes out on the A915. Gates across the way lead to Largo House, now derelict but with the fine Adam front still intact. Turn right, up the hill, to Upper Largo. The inn sign depicting the *Yellow Carvel* commemorates one of Sir Andrew Wood's ships (Sir Andrew was a 15th-century naval commander, a native of Largo, noted for his fights with English ships in the Forth). Cross and take the first road left

INFORMATION

Distance: 13km (8 miles). Up and down Largo Law only is 6km (4 miles).

Start and finish: Lower Largo, Temple carpark.

Terrain: Mostly on roads and tracks, but on the Law, with its very steep hillside, boots are advisable.

Toilets: Temple carpark.

Refreshments: Lower Largo.

Note: Dogs are not allowed on Largo Law.

Lundin Links standing stones looking towards Largo Law.

The historic church and Largo Law.

(just past the garage), which dips down and then heads off over the west flank of Largo Law.

Before heading for the hill, divert to the church, off left on its commanding site, to see a Pictish symbol stone just inside the gateway and, on the south side, a stone with elongated figures on one side and their heart-rending story on the back. Sir Andrew Wood is buried in the church.

Turn right coming out of the churchyard then take the first road left, which will take you back on to the hill road. Note the 'piebald' cottage, quite a local feature and often the black is simply the dark whinstone. Peer over the wall on the left before going further and you'll see a slight depression curving away towards a small white tower with a conical roof – all that is left of Sir Andrew Wood's castle. The depression marks a canal (the first ever cut in Scotland) which he had made by English prisoners so he could go to church in his admiral's barge – or so the story goes.

Continuing up the hill road, you come to the primary school and modern graveyard. Between them, a kissing gate is the start of the recognised route up Largo Law.

An edge of a field and then a track leads up to Chesterstone Farm, passing a row of old cottages neatly converted into a house. Follow the track round the farm till, left, there is an obvious green lane heading straight for the Law.

The final ascent is one of the steepest of all the walks described in this book and whin bushes allow no deviations low down. You arrive on a fine summit only to find there's a dip (where there is a fence with a stile) and the real summit beyond, with a cairn and trig point.

The view is extensive: the sweep of Largo Bay along to the sprawl of Leven and the far Binn with its relay mast; seaward lie the Isle of May and Bass Rock; while the Lomonds, Ochils, Sidlaws and, on the horizon, the Highland hills, all give a variety of mountain scenery. The summit area is surprisingly grassy and full of yellow violets, bedstraw, speedwell, buttercups and cuckoo flower.

The original edition of this guide gave a return circular walk via Lahill Craig; this – as notices indicate – is no longer practical, so you can either retrace the outward route or, better still, enjoy the walk by taking in the Woodland Trust's Keil's Den, which is rich in birdlife. This walk is perhaps at its finest in May when the massed wild hyacinths (English bluebells) and ramsons are in flower. Also at this time, the rich landscape is patched with vivid, yellow fields of oilseed rape.

Descend through Chesterstone Farm and back to the school and cemetery then turn up the tarred road. Just beyond the cemetery entrance, head off left on a path that follows the field edge. This path crosses a strip of wood and another field to exit on to a minor road by the Woodlands Garden Camping and Caravan Park. Walking over, you are treated to a closer view of Andrew Wood's tower and there is also a lectern-style doocot. Cross the road and go between fields to reach Keil's Den ('den' is Scots for 'dell').

Turn left along the east bank of the den and, at an area of tree replanting, you can take the path that drops back right to cross and follow the west side of the den, a slightly longer option than continuing on the east side.

If continuing on the east side, the path descends and swings left to reach the minor road at Largo House Caravan Park. (Where the path swings left, another path angles back down to cross the burn to join the west path at stepping-stones.) Turn right at the tarred road that leads down to the A915 coast road. Harbour Wynd, opposite, leads back to the sea at Lower Largo, where you turn left along to the start.

This route leads you past Alexander Selkirk's cottage. At the A915, turn right and after a few minutes you reach the Cottage Tearoom and Restaurant, which offers a warm welcome and good food and is open all year.

If you have opted to walk down the west side, the path eventually leaves the woods and

crosses cultivated land to reach another tiny road. Turn left here, crossing the Keil Burn – and you soon reach the A915. (Please note that 'Lundin Mill' is a housing estate and not an actual mill.) Turn left along by the Cottage

'Robinson Crusoe': the Alexander Selkirk memorial, Lower Largo.

Tearoom then right on to the Harbour Wynd
and then home. One of Fife's decorative
milestones may be spotted at the A915/
Harbour Wynd corner.

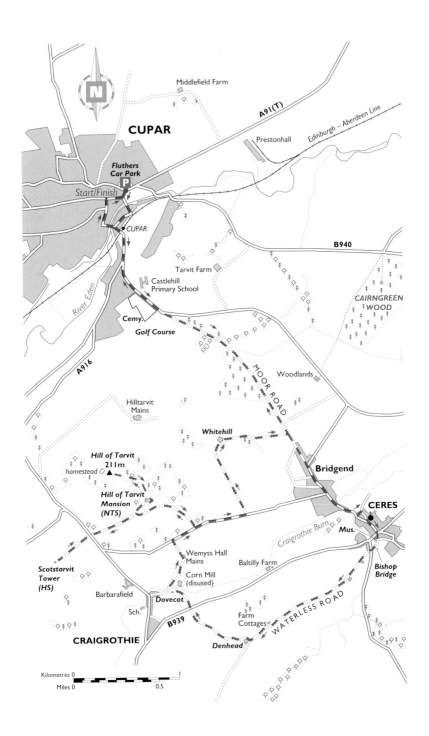

Middlefield Farm

A91(T)

Edinburgh – Aberdeen Line

CUPAR

Prestonhall

**Fluthers
Car Park**

Start/Finish

CUPAR

B940

Tarvit Farm

Castlehill
Primary School

CAIRNGREEN
WOOD

River Eden

Cemy.

Golf Course

A916

Woodlands

MOOR ROAD

Hilltarvit
Mains

Whitehill

**Hill of Tarvit
211m**
homestead

Bridgend

CERES

**Hill of Tarvit
Mansion
(NTS)**

Craigrothie Burn

Mus.

**Scotstarvit
Tower
(HS)**

Wemyss Hall
Mains

Baltilly Farm

**Bishop
Bridge**

Barbarafield

Corn Mill
(disused)

WATERLESS ROAD

Sch.

Dovecot

Farm
Cottages

B939

CRAIGROTHIE

Denhead

Kilometres 0   1
Miles 0   0.5

THE MOOR ROAD, CERES AND HILL OF TARVIT

This is best done as an afternoon walk as the main visitor attractions are only open then. It is not really suitable for dogs or very small children.

Take care crossing the busy main road on leaving the carpark then turn right and skirt the ornate war memorial. Fork left and carry on round to a junction (Cupar station is on the left). Turn left, over the Edinburgh–Aberdeen line (the statue seen here is of one of the line's benefactors) and walk on out of Cupar awhile before turning left on to the Ceres road. The town ends with the cemetery and golf course, and then there is an obvious kissing gate into trees and a footpath sign for Ceres.

The Moor Road is typical of older routes in the pre-turnpike days when boggy valley bottoms were avoided. The shady climb is pleasant, there is some open ground on top and then the path runs down to meet the main road at Bridgend, as this part of Ceres is called. Continue down into the village, an old, trim place, and opposite Meldrum's Hotel, turn up Curling Pond on to Kirk Brae and visit the Griselda Hill Pottery – this specialises in recreating the once-famous Wemyss ware and is now highly collectable in its own right. You can see the pre-glaze painting in progress from the showroom.

A little further on is the 1806 church, a large sandstone block with a landmark steeple. The site is ancient and the graveyard has some old stones and a flag-roofed mausoleum dated 1616. See if you can find the wall monument to Rev. J.C.C. Brown and the heartbreaking story it tells.

Drop down to rejoin Main Street (the village's shops) and on to a crossroads by the old coaching inn. Opposite is the toby jug figure of The Provost, which is actually the depiction of a local divine.

INFORMATION

Distance: 13km (8 miles).

Start and finish: Cupar, Fluthers carpark, east end of town.

Terrain: Road, tracks and clear paths, with grassy slopes on Hill of Tarvit. No special footwear required.

Toilets: Several in Cupar, including Fluthers carpark, Ceres carpark, Hill of Tarvit mansion.

Refreshments: Coffee shop next to Ceres Museum; Hill of Tarvit (when open); wide choice in Cupar.

Opening hours: *Folk Museum:* Easter and mid-May–early October, daily, 1400–1700. *Hill of Tarvit Mansion House* (NTS), Easter and May–September, daily, 1330–1730; October, weekends only, 1330–1730. *Gardens:* all year, 0930–sunset. *Scotstarvit Tower:* as for Hill of Tarvit Mansion, where keys are issued to visitors. *Griselda Hill Pottery:* Monday–Friday, 0900–1630; Saturday, Sunday, 1400–1700.

Tourist office: St Andrews, tel: 01334 472021

Weigh-house, Ceres.

Cross to the High Street, and next to Brands Inn is the Fife Folk Museum in the 1673 Weigh-house (note the plaque over the door), which once served as the village tolbooth and even had a dungeon. On the wall outside are the jougs: an iron collar for holding wrongdoers. The award-winning museum has a vast collection relating to life in past centuries and the shop in the annexe across the road has local-interest booklets for sale.

Beyond the museum, walk down the lane and over the cobbled 17th-century Bishop's Bridge. This leads to a carpark, toilets and the Bow Butts where the 700-year-old free games (incorporating the Ceres Derby) are still held on the last Saturday in June. There's a monument to the Battle of Bannockburn, in which Ceres bowmen took part (interpretative boards).

At the south end of the carpark, Woodburn Road leads off left of the row of cottages, a quiet lane historically called the Waterless Road, as it used to be the road from St Andrews to Pettycur and the Edinburgh ferry. Archbishop Sharp came this way shortly before his assassination by Covenanters on Magus Moor in May 1679. As you climb through the fields, Hill of Tarvit rises clear. The monument on top is to commemorate Queen Victoria's diamond jubilee. Shortly after a row of cottages, you reach Denhead Farm. Turn right around the walled-off house to follow the drive out to a tarred road. This is a private drive not a right-of-way, and is most attractive. Cross the road for a continuation of the path. When this comes out on a track, turn left, passing a derelict lectern doocot. Turn right at the next junction and down to a ford/arched packhorse bridge then up again to reach a road. Turn right, and a few minutes' walk leads to the east drive of Hill of Tarvit Mansion. Enter here (cars are directed to the west drive) and walk up to the house, skirting a big yew hedge.

The house is an NTS showpiece, notable for its furniture, the kitchen, Chinese porcelain, paintings (British/Dutch) and the gardens. There is also a welcome café with home-baking. From the top of the walled garden, imposing gates give access through woods to the open hillside above. Angle up left (the slope is steep and grassy) to the top of the actual Hill of Tarvit, another of Fife's panoramic viewpoints.

The mansion house was built by Sir William Bruce to supersede Scotstarvit Tower. Originally called Wemysshall when owned by the Wemyss family, it was sold to the Dundee jute magnate, Frederick Sharp, whose daughter bequeathed it to the NTS, along with its treasures. Sir Robert Lorimer reshaped the house and laid out the gardens in 1906.

Hill of Tarvit Mansion and the monument on the hill.

Scotstarvit, a typical laird's tower house, lies 1km west of the mansion and, if requested, a key can be taken from reception to visit it. Originally built in the 1500s it was bought in 1611 by Sir John Scott, the cartographer (he employed Timothy Pont to map Scotland) and literary enthusiast (his brother-in-law was Drummond of Hawthornden). He married three times and had 19 children.

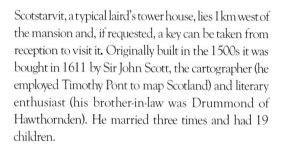

Leave the mansion by the east drive again, turning left then, after 500m, left again up the drive to Whitehill. Pass the house and skirt right of the farm, pass the cottages and the edge of a conifer stand, enjoying fantastic views right up to the Angus coast. Go through the right-hand side of two gates and take a path down the field edge to rejoin the Moor Road – 'He tae Cupar mun tae Cupar.'

Once in the town, don't turn right at the station but walk on to a T-junction and turn right into the Crossgate. Pass the post office and spired library building (local booklets on sale), and at the end is the Mercat Cross and the clock tower of the old town hall. Left lies the Bonnygate, with plenty more shops and cafés, while right is St Catherine's Street, with the old Corn Exchange building, which leads back to the starting point.

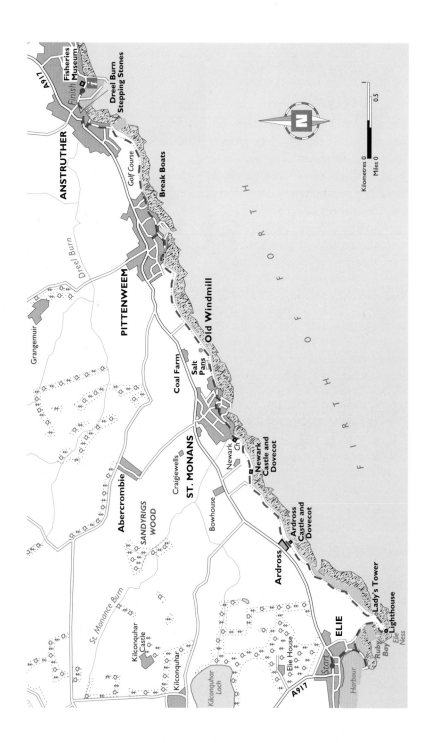

EAST NEUK VILLAGES

This walk constantly changes from town to coast to town, so there is plenty to see and do and plenty of chances for refreshment *en route*. Walk down to the sea at Elie and out first to the tidal harbour and the Granary building before heading off along the coast, passing through a carparking area overlooking Ruby Bay (the rubies are only garnets!). The coastal path is signposted. In 1908 David Stevenson built Elie Point's lighthouse buildings. The next tower was built c.1760 as a summerhouse for Lady Janet Anstruther, as was the grotto in the back of the bay below. Here her ladyship came to swim – and a bellman was sent round the streets of Elie to warn the plebs to keep away. Dunes lead on towards Ardross, a castle in very ruinous state, though inland, by the farm, is a fine example of a lectern-style doocot.

The path struggles along through geological chaos to reach Newark Castle, a weathered ruin, which was the home of David Leslie, the general who defeated Montrose at Philiphaugh, near Selkirk, in 1645. Walk on from the beehive doocot and take the paths that pass below the church – which is worth a visit. It is a well-known seamark and has been tastefully restored. Note the models of old ships hanging up inside. Cross the burn and go right of the house opposite and climb to a road then turn down again to St Monans (St Monance) West Shore and harbour.

INFORMATION

Distance: 9km (5½ miles).

Start and finish: Elie and Anstruther are linked by coastal bus services (hourly) and parking is available in Elie. For transport information, tel: 01592 416060.

Terrain: Coastal paths and towns. Stout shoes advised.

Toilets: In the towns.

Refreshments: In the towns.

Opening hours: *Scottish Fisheries Museum, Anstruther:* April–October, 1000–1730, Sunday, 1100–1700; November–March, 1000–1630, Sunday, 1400–1630. *Windmill:* Office hours in July–August, or key from newsagent in Forth Street, St Monans.

Tourist Information: Beside the Scottish Fisheries Museum, April–October, tel: 01333 311073.

Elie.

St Monans.

St Monans is a delightful place and it is worth walking round some of the old streets facing the harbour to see the typical houses, many restored under the NTS Little Houses scheme. When heading east again do not go along East Shore (which looks like the route) but up and then right, along Rose Street.

Pass an abandoned swimming-pool and an area where salt was once produced. Coal Farm is a reminder of that era. It took 32 tons of water and 16 of coal to make one ton of salt. The windmill is the old pump for these workings (interpretative boards).

Some open country follows then you climb to a children's play area and, suddenly, there is a view of Pittenweem's West Shore with the houses rimming the bay. Descend by the path behind the shelter and go along beside the houses. Pittenweem is now the main active fishing port on this coast with its market and other facilities. At the end of the harbour lie the much-photographed houses of The Gyles – used in the film *The Winter Guest*.

The High Street *is* high and you can reach it by Cove Wynd – which should be Cave Wynd after St Fillan's Cave, which is passed. (St Fillan was a 7th-century missionary.) The key for the cave is held at the Gingerbread House (café/crafts) on the High Street. The cave – with its well – is still used as a chapel, and the underground stairs led up to an abbey at one time. The Kellie Lodging on the High Street is a 16th-century laird's town house, and the parish church tower is the old tolbooth. Pittenweem explored and yourself refreshed, push on for Anstruther.

Pittenweem – Water Wynd.

From The Gyles, head up Abbey Wall Road and take an opening to pass a play area on the cliff top. The striated reefs below are called, ominously, the Break Boats. The cliff is rather breaking, too, but follow a path down to the shore again and carry on by the edge of Anstruther golf course, along by the shore to Billow Ness and a last bay where fences protect walkers from sliced drives. The battlemented tower on the golf course is the war memorial.

Follow the curve of Shore Road then right as far as the school then left up to the main road. Turn right on to the High Street, passing the Dreel Tavern. The road makes a sharp bend, left, to bridge the Dreel Burn but, after noticing the Buckie (Shell) House on the corner, cut down a small road on the seaward side to cross the Dreel Burn by stepping stones – if the tide is out. (At full tide you will have to return to the main road.) Plaques here and elsewhere show connections with the era of clipper ships, with locals owning or captaining some famous names like *Ariel*, *Taeping*, *Min* and *Lahloo*.

Anstruther Harbour is the largest of the East Neuk ports but no longer used by the fishing fleet. Facing the harbour at the far end is the famous Scottish Fisheries Museum, the best of its kind anywhere, and it also houses a tearoom. Anstruther is locally pronounced more like 'Enster', so don't be confused if you hear this.

Allow a good hour for the museum – and refreshments – before heading up to the main road for a bus back to Elie. Information from St Andrews Bus Station, tel: (01334) 474238, or ask locally (Tourist Information Office beside the Fisheries Museum), or use a taxi.

Anstruther. When there, see if you can spot what has changed in the view.

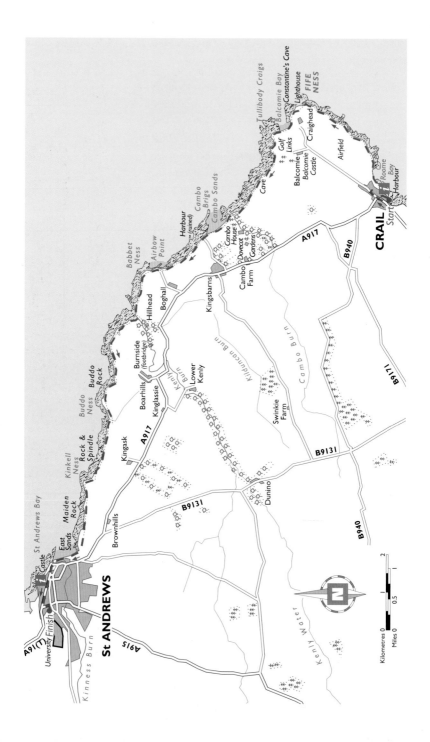

CRAIL TO ST ANDREWS COAST WALK

This is the longest, hardest and loneliest walk in the book, and is only for the fit and experienced. In reality you can hardly go astray: the sea is always on the right! One river could be paddled (or a long detour made) and there is nowhere offering refreshments. Don't drink stream water. High tides can block several sections.

Crail is a showpiece of vernacular architecture, its tiny harbour a gem, the tolbooth showing Dutch influences (local museum beside it); from there head east along the Marketgate. At the far end is the parish church with a unique collection of early mural monuments, and, in season, the graveyard and woods are a mass of snowdrops. Legend has it that the stone at the entrance gates was thrown at the church from the Isle of May by the Devil. He missed and the stone split, one part falling here, the other on the coast at Balcomie. (These are erratic blocks, left by glaciers, often called 'blue stones' and given superstitious origins.)

Turn seawards along Kirk Wynd and down a footpath past the 'pepperpot' doocot, which is painted white as a seamark. Turn left round Roome Bay, pass the eroded red cliffs beyond (coastal path sign) and join the road through the Sauchope Links Caravan Park. Continue on at shore level by a knobbly 'castle' rock and the tangly Kilminning Nature Reserve (Scottish Wildlife Trust). Above, stretching from the old airfield to Fife Ness, is the new Craighead Links golf course. A coastguard station and lighthouse mark Fife Ness, and there is a private bird hide and an interpretation board (Beacon base template, 1813). The North Carr lightship was originally positioned off the point, which is an excellent place for observing migrant seabirds, including gannets – the Concorde of the seas.

Follow a small, tarred road to the trim Balcomie golf course (founded in 1786), whose edge you go around,

INFORMATION

Distance: 22km (14 miles).

Start and finish: Crail or St Andrews, but the best choice is to drive to St Andrews (street parking and carpark) then bus or taxi to Crail for the start. St Andrews bus station, tel: 01334 484238, or phone: 01592 416060. Time-tables are on display.

Terrain: Rough coastal walking: proper walking boots and equipment essential.

Toilets: St Andrews, Crail, Cambo Sands carpark.

Refreshments: Only at start and finish. Always carry adequate food and drink, suitable for a long, hard walk.

Opening hours: *Crail Museum:* April–May, weekends and holidays, 1400–1700; June–mid-September, Monday–Saturday, 1000–1300, 1400–1700, Sunday, 1400–1700. *Cambo Gardens:* 1000–1600.

Tourist Information: St Andrews, tel: 01334 472021; Crail (April–September), tel: 01333 450869.

Crail Harbour.

close to the shore. (The gritty track leads past Constantine's Cave, named after the king who was killed there by the Danes, c. AD 874.) Sandy Balcomie Bay is home to the old RNLI lifeboat building and, at the far end, the second of the blue stones mentioned earlier.

Crags run down into the sea and may be impassable at high tide. There are no easy diversions. Beyond, is a new golf course which stretches beyond Cambo Sands. The estate has signposted any practical stretches of path, and some parts have Fife Coastal Path indicators.

The Cambo Glen burn has a substantial bridge, while streamside paths (edged by snowdrops in spring) lead to the 'Secret Garden' – open to visitors – a view of Cambo house and a fine doocot, a diversion that is recommended. Back at the shore again take the path on to the warm-coloured Cambo Sands (swimming dangerous). There's a carpark/picnic area, the only public approach to the coast – a road from Kingsbarns.

Very rural walking follows. Beyond Babbet Ness the coast becomes a ragged chaos, this jagged rock being sandstone which was laid down in rivers, as opposed to the smoother sandstone from lake sediments. Partridges may go shooting off and make walkers jump. Farther on, at an area of sea walls, there are fossil imprints of prehistoric roots. The decaying building is an old salmon fishing bothy.

Keep along the track outside the walled field to reach the Kenly Water – which has to be crossed. At full ebb it is easy but, if you decide to paddle, choose the wide, gravelly area (facing the slip) rather than the rocks further upstream – they can be disastrously slippery. The nearest footbridge is at Burnside Farm, a safer diversion, which is very pleasant in itself, and is the Fife Coastal Path recommendation as the shore route is difficult once over the Kenly Water.

Walk up past Burnside Farm, turn left on the minor road, follow it up a rise and, when it swings left, take the track across the fields, right. Turn left then right, to round some big, green barns – note the pantiled doocot – then break off, first on the right, on a green track back to the shore.

Sandstone country (red and grey) is reached at the Buddo Rock 'castle', which has a natural arch and a central passage that allows a scramble to the top. Angle up to follow the top of the bank – where there is a wall – to the viewpoint of Buddo Ness then down again to traverse the loneliest section of all: a switchback path through thick scrub, which leads to Kittock's Den, another glen running up inland. Except at high tide you walk on the shore, cross a wall and, beyond, reach another volcanic area. Paths are being constructed along an overgrown bluff where the ground is rather eroded. This thorny section leads on to easier ground and along to the Rock and Spindle, a phallic-like pinnacle.

Beyond is Kinkell Ness and, at some steps, head up on to the heights as the shore becomes almost impossible to negotiate. St Andrews is suddenly near and you come across the sprawl of a caravan park, which you skirt around before dropping to the East Sands. (Below, another sandstone pinnacle is the Maiden Rock.)

The East Sands lead on past a leisure centre, coastguard building, the Gatty Marine Laboratory and a putting green to reach the harbour, which is crossed by a retractable bridge. St Rule's Tower and cathedral ruins, above, show the route into town. You arrive at North Street which is followed to the far end. At the roundabout, turn left, up the hill, and you come to the Bus Station.

St Andrews Harbour.

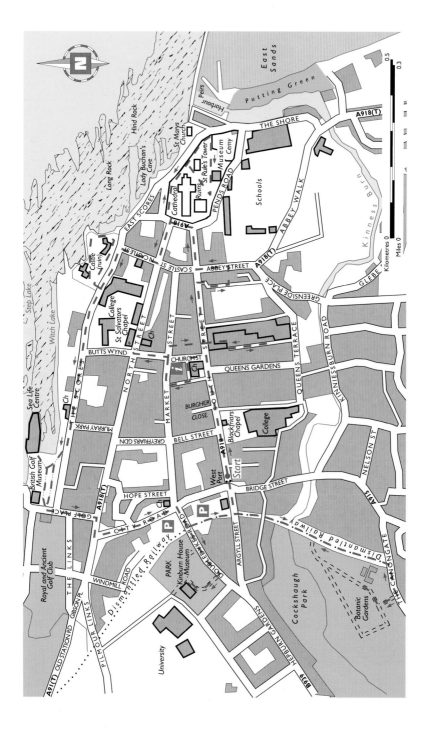

EXPLORING ST ANDREWS

This urban day is quite tiring if most of the sites are visited. Start at the West Port, the old walled gateway to the town. Walk along attractive South Street to the isolated ruin of Blackfriars Chapel with the lawns of Madras College behind. A notice describes the school, founded by a missionary to India, its 'suave Jacobean manor' appearance hiding early austerity. (One rector wrote of donning two sets of woollen underwear in October and keeping them on till April.)

Continue along South Street to the Town Hall with its distinctive clock suspended over the street. The wartime Polish forces are commemorated by a mosaic mural. Now cross over to the Town Church.

The Town Church (Holy Trinity) is historic but may not be open. The huge mural tomb to Archbishop Beaton, murdered in the castle in 1546, is a notable feature. John Knox preached an epoch-making sermon here in 1559. The chancel floor is of Iona marble, the roof is of Caithness slabs.

Go through Logie's Lane, behind the church, and right on to Market Street, once the heart of the town with cross and tolbooth, and still observing the ancient Lammas Fair. It was also the site of executions: of the hapless Chastelard (see Walk 8) and of Paul Craw, a Bohemian reformer burned in 1433. Cross over and go along College Street to reach North Street, dominated by St Salvator's Tower, a splendid medieval example. On the pavement by the pend (passage) the initials P.H. are for Patrick Hamilton, the first Reformation martyr, burned here in 1528. Through the passage, right, is the church entrance. The university sprawls all along North Street, and only Oxford and Cambridge are older foundations.

Return to Market Street, cross leftwards by the Tourist Information Centre and follow Crails Lane through to South Street. Opposite is the arched entrance

INFORMATION

Distance: 4km (2½ miles).

Start and finish: Anywhere in the town. Avoid expensive parking spaces on the main streets. Carparks are signposted. It is worth picking up a town map from the Tourist Information Centre, 70 Market Street, tel: 01334 472021, before commencing the walk.

Terrain: Streets and roads, ordinary shoes adequate. St Andrews can be windy and cold, so be clothed suitably.

Toilets/Refreshments: Available throughout the town and at the museums.

Opening hours: The Cathedral (Museum) and Castle (Visitor Centre) offer a combined ticket; St Rule's Tower: entrance tokens available from the museum: April–September, daily, 0930–1830; October–March, Monday–Saturday, 0930–1630, Sunday, 1400-1430. Preservation Trust Museum: June–September, daily, 1400–1730. Golf Museum: Easter–mid-October, daily 0930–1730; rest of the year, Thursday–Monday, 1100–1500. St Andrews Museum: April–September, daily, 1100–1700; October–March, Monday–Friday, 1030–1600, Saturday and Sunday, 1230–1700. Botanic Gardens: open 7 days a week, October–April, 1000–1600; May–September, 1000–1900; glasshouses, October–April, weekdays only, 1000–1600.

of St Mary's College. The quadrangle has a holm oak planted in 1728 and a centuries older thorn. Explore the gardens and, on leaving, turn right along South Street.

Turn into South Court Pend (Byre Theatre sign) where there's a plaque to J.D. Forbes, the alpinist who discovered that glaciers were moving rivers of ice. Byre means 'cowshed' and the tiny original theatre was created in a cattle shed in 1933 (it is now in a new building). As you near the end of South Street, the cathedral ruins appear framed between the (leaning) Roundel and Queen Mary's House. The Pends' gateway led to an old priory and down it, right, is St Leonard's School, a building that began as a 12th-century hospice.

Facing The Pends is Deans Court, once the Archdeacon's residence. The cobbled symbol marks where 80-year-old Walter Myln was burned. Go through the gate into the cathedral grounds; once the second largest in Britain, it is now a 'rent skeleton' (Ruskin) and the main interests are St Rule's Tower (small charge for climbing 33m to the viewpoint on top), the museum and wandering round to look at the monumental stones, which range from full-rigged ships to golf champions, including young Tom Morris who won four Opens in a row before dying, aged only 24. The cathedral suffered during the Reformation and its stonework was pillaged for town buildings. The museum has a collection of early stones, 'green man' carvings, a skeleton stabbing a victim in the back and, beautiful beyond description, an old sarcophagus panel alive with the animals of a hunting scene.

The view from the top of St Rule's Tower.

Leaving the grounds, walk along North Street. The Preservation Trust's Museum is on the left. Its mock-up of a Victorian shop is entertaining. Turn right into North Castle Street and ahead looms the castle itself. The G.W. initials outside are of the reforming preacher George Wishart, who was burnt here in 1546 while Cardinal Beaton watched from a castle window – out of which his body was soon to hang. Unusual are the mine and counter-mine (which you can explore) and the notorious bottle dungeon (which is not recommended!). The 1546 mine was cut through solid rock to place charges under the castle's walls to blow them up but the defenders burrowed a counter-mine and intercepted the attempt. There is a good visitor centre and shop.

Turn right on leaving the castle to walk along The Scores to reach the obelisk of the Martyrs Monument. Below it are two modern attractions. The Aquarium is a wonderful window into the sea world (and the balcony café over the sea may be very welcome) while the British Golf Museum uses every high-tech device to captivate anyone even slightly interested in golf. The building opposite is the Royal and Ancient clubhouse with the Old Course beyond – the very home of the sport.

The West Port can be regained from the museum by walking along Golf Place, turning right then first left along past the Bus Station. If legs and brain can stand any more, off Doubledykes Road (right at the roundabout beyond the Bus Station) is St Andrews Museum which uses modern techniques so you can see, hear and even *smell* the past. From the West Port if you head off along Argyle Street (the opposite direction to the morning's start) you soon see a sign indicating the Botanic Garden. This is a ten-minute walk away and, besides the attractive layout, has a cactus collection (250 species) and an orchid house.

You may well have to be somewhat selective with the above. If less interested in the historical sites, the museums are fun, and from the Golf Museum you are close to the huge West Sands, which are a favourite walk for 'town and gown'. The opening shot of the film *Chariots of Fire* showed the athletes running along the tideline here. The famous golf courses lie inland from the sands. There is also the ladies' putting green (open to the public) which is known locally as the 'Himalayas'.

Royal & Ancient clubhouse.

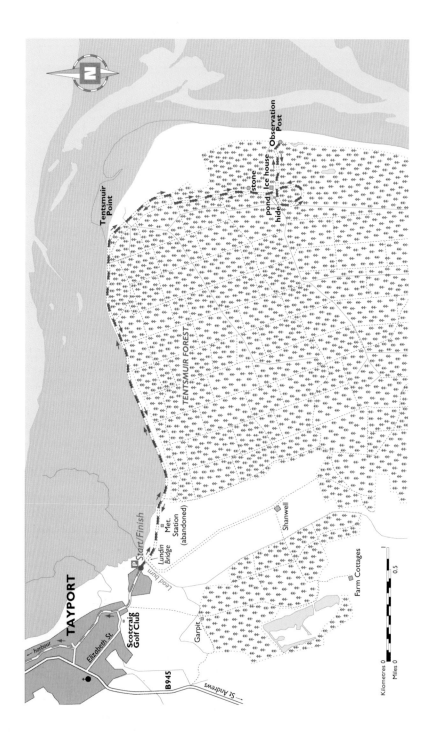

TAYPORT AND TENTSMUIR FOREST

Tentsmuir is a huge forest, created in 1922 on what were basically salt marshes at the north-east corner of Fife. Like most forests, the main interest and wildlife are found round the edges and open areas, so this walk keeps to the coast in both directions. (Circuits through the forest interior are a dull option, navigation can be difficult and felling interrupts progress.) Some of the area is a National Nature Reserve, of which more later. Take a picnic and plenty of liquid, and go early (and quietly) to maximise the chance of seeing wildlife. Most of the trees you'll see are Scots pine.

The easiest approach to the start is from the St Andrews road. On entering Tayport the striking white church of Our Lady Star of the Sea is seen, left, and then, right, Elizabeth Street is indicated with a sign for the golf club. Head down to pass the Scotscraig Golf Club and, at a T-junction, turn right, into Shanwell Road, and follow this to its end. Park on the edge of the grassy area looking on to the sea and Tentsmuir Point.

Head off around the coast from Lundin Bridge. The tiny Lead Burn hardly seems sufficient to have powered mills last century. Keep outside the old Meteorological Station then follow the track in to the corner of the forest. The shore is barred by concrete blocks laid as anti-tank defences in the 1939–1945 war when invasion was a possibility. There's a confusing number of tracks but turn left (almost at once) at an open space to walk along just outside the permanent forest on a good track (trees on the left are self-seeded). Gaps to the left allow views over to the Angus coast. The sands at low tide can be 1km wide.

There's one 100m stretch where the sea is winning over the land and the pines are tumbling from erosion. The path has gone, so cut along through the trees or walk the sand, depending on the tide. (The concrete blocks have

INFORMATION

Distance: 13km (8 miles).

Start and finish: Lundin Bridge, south of Tayport, reached via the golf course approach road.

Terrain: Sandy paths and tracks in forest. No special footwear needed.

Toilets: Tayport.

Refreshments: Tayport.

The drive to Kinshaldy.

been tossed about by the storms too.) The next landmark is a gas pipeline marker then a drain, and finally the point is reached at the National Nature Reserve (marked by a fence and with an information board). Cross the stile, take the path on top of the dune to the right and follow it towards then beside, the concrete blocks which marked the edge of the sea during World War II. Eastward, the sand is still building up steadily as the River Tay brings down sediments. (Some of the grains you walk on originated on Schiehallion or Ben Alder.) Scientists are studying this natural reclamation, and all the land east of the fence is a reserve (SNH).

Keep by the blocks until the path meets a vehicle track. Turn right, towards the forest take the track out of the reserve (stile, notice-board) and go on a short distance until you reach a T-junction. Turn left along this bigger track, passing fenced-off, open areas. About 100m past this open space, keep an eye open for a stone like an elongated milestone (left). The inscription shows it to be a 1794 boundary stone between two salmon fisheries – and it is aligned with the Ochils' peak of Norman's Law. It once stood at the edge of the shore.

Next there's a fork, with one branch of track turning inland, but keep straight on and in a few minutes reach your destination: the large icehouse on the left. Icehouses were half-buried and enclosed like this in order to keep the ice from melting in the days before refrigeration was available to fishermen. Ice could even be carted from the Highlands. This must have been a remote and desolate spot in those days.

The old icehouse in Tentsmuir Forest (the lintel is dated 1888).

Just beyond the icehouse area, a small footpath goes off right (marker posts) and leads to a pond much favoured by dragonflies. Tentsmuir has very diverse bird life; you may see herons, moorhens and gulls here, as well as wrens, green and great spotted woodpeckers, sparrow hawks, curlews and warblers. Other species abound: colourful snails, rabbits galore, foxes, roe deer and red squirrels. Go quietly and you should see most of these. Bat boxes have encouraged three species to live in the forest.

The pond in the heart of Tentsmuir Forest.

Backtrack to the fork that led to the pond. Follow the other fork thereafter. It takes the line of a beech avenue then swings on to the main forestry track. Turn left and, after about 500m, you are back at the icehouse. On the seaward side there's a track going to a gate/stile with a reserve sign. Take this and walk out to the dunes at an old wartime lookout post. The concrete blocks run along just in front of it, quite invisible as they are under the dunes.

This is the place to picnic. Seals haul out on the distant sands. There are dangerous currents, so do not swim or even go out any distance from the tideline. If dunes fascinate, a separate visit is recommended to Kinshaldy at the south end of Tentsmuir Forest. It is well signposted from the Tayport–St Andrews road, and there are marked trails, picnic and barbecue areas and toilets. You now wander back to Tayport.

Leaving from where you parked, drive straight on instead of turning left to the golf club. This becomes Nelson Street. At its top end, take the second right which leads along to the attractive harbour area (Bell Rock Tavern is a landmark) where there are toilets by the carpark. Tayport (Ferry-Port-on-Craig) was for a long time the great crossing point for the Tay. Broughty Ferry, on the other side, is still marked by a castle (museum). The Tay Bridge ended the ferry days. Broad Street/Castle Street leads up and is the main shopping street and, turning right at the end, points you along to the Tay Bridge and wherever. A rude old saying is: 'Out of the world and into Fife.' For Tentsmuir it is rather apt.

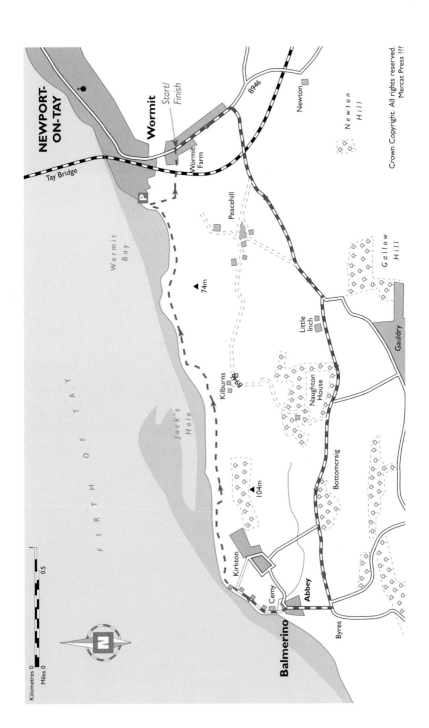

NEWPORT-ON-TAY

Tay Bridge

Wormit

Start/Finish

Wormit Farm

P

B946

Newton

Newton Hill

Peacehill

▲ 74m

Wormit Bay

Jock's Hole

F I R T H   O F   T A Y

Little Inch

Kilburns

Naughton House

Bottomcraig

Gallow Hill

Gauldry

▲ 104m

Kirkton

Cemy

Abbey

Byres

Balmerino

Kilometres 0
Miles 0
0.5

N

BALMERINO AND THE TAY ESTUARY

Pause when driving into Wormit to enjoy a close view of the Tay Bridge. (There is a parking area near the signal box, which was once Wormit Station.) The bridge, of course, is the second to be built over the Tay. The first bridge collapsed in a storm in December 1879 taking a train with it at the cost of 75 lives. The stumps of its pillars can be seen just east of the present 85 span, 3½km sweep.

Wormit grew up as a dormitory town for Dundee following the establishment of the railway bridge. It was the first town to be lit by electricity and has plenty of 'assured' Victorian buildings.

Driving on into Wormit note the 'extravaganza in black and white' shop-on-the-corner building (right) then turn left on to Bay Road and follow it to its end. There is parking by the shore.

Take the path up the edge of the field by the last house, pass a pond at the foot of Wormit Den and then go left, passing underneath the railway line on a track to reach the B946 road. Turn right on this road and follow it as it climbs steadily out of Wormit. There is an expanding view across the estuary to Dundee.

Take the first turning on the right (signposted to Gauldry and Balmerino). Walk along this quiet road, which is also part of the extensive network of Fife Millennium Cycleways. Cross the railway and pass the entrance to Peacehill Farm. There is an element of black humour about the name, for a gallows once stood here.

Continue along the road, and at the fork keep right (the left fork goes to Gauldry). This corner of Fife is still intensively farmed and grows vegetables including early potatoes, lettuce and broccoli, as well as various cereal crops. Animal-wise, there are also beef cattle, and at Peacehill you will also see large poultry sheds.

INFORMATION

Distance: 10km (14 miles).

Start and finish: Wormit, on the Tay Estuary, facing Dundee. Parking by sea, west end of town.

Terrain: Road, tracks and path. Some muddy areas. Boots recommended.

Toilets: Wormit and Newport, also by Tay Road Bridge carpark.

Refreshments: Limited choice in Wormit and Newport. Wide range in Dundee.

Opening hours: *Balmerino Abbey*: open all year, free (NTS).

On the Balmerino–Wormit coastal path.

The road passes through Bottomcraig, and it is worth noting that although Kirkton of Balmerino is down by the sea and is passed later in the walk, however, the church or kirk for the area is now up here. Ignore the right turn signposted for Kirkton and continue for a further 500m to a complex junction. Turn right here for Balmerino (locally pronounced Bamérnie).

As you walk downhill you come to an unusually formal square on the right, which is a war memorial dating from 1948. The story is told in the Doric-pillared portico. The row of statues depict the four seasons. The cottages are lived in: a very practical memorial.

Continue a little way further to reach Balmerino Abbey, also on the right. The Abbey (NTS) dates its foundation back to 1226, but was burnt, like so many of its kind, by English raiders in 1547 – and pillaged again after the Reformation. The vaulted sacristy and the chapter house survive, but the most unusual object is a Spanish chestnut tree, which is believed to be 400 years old. Stone for the Abbey was brought by sledge and water from Strathkinness, near St Andrews.

The walk continues down the 'no through road' lane ahead, but you might like to divert right along a narrow lane to the cemetery, which has some interesting stones. On one, the lettering is upside down, while another has an unusual crook symbol, showing that it commemorates a shepherd.

Walk down the lane to the shore, passing a beautifully restored mill building. The white cottages further on are converted fishing bothies, and as Nether Kirkton is reached there is a large, round erratic boulder on the beach, called Samson's Stone. A sign shows that the footpath runs between the last houses and the shore, but once past the buildings, you follow white-topped marker posts to the top of a bank where the path (which can be narrow and muddy) continues through trees before emerging on to open ground at a signpost. Scurr Hill rises to the right.

The path continues along the field edges, which are generally fairly clear but again often muddy in places. Below Kilburns there is a deep, wooded den to cross, then the view opens out. Go through a gap in the fence and follow cattle paths along. Whin gives way to dog-rose and elder as the summer advances, and rape fields can be scented from miles off.

As you near Wormit Bay, keep above a dense area of gorse then follow the fence until the path dives down a buckthorn tunnel to reach sea level. Walk around the bay enjoying the expansive views and the birdlife. The rail bridge sweeping across to Dundee is very impressive from here.

Back at the start, you should not miss driving east towards the 1966 road bridge, both for the views and to enjoy the swanking villa architecture of the bold Victorian jute barons who built their houses here – well away from their smoky factories in Dundee. Woodhaven was an early ferry point but lost out to Newport. Dr Johnson crossed from there and, in his journal, complained at the fare.

Drinking fountain at Newport.

Cow's Hole and Pluck the Crow Point lead to Newport-on-Tay, with no break in the buildings, but the town centre is marked by several shops and, off Cupar Road, a cluster of churches. Don't miss the 1882 canopied drinking fountain above the sea: wrought iron at its most elaborate. Thomas Telford was the engineer of Newport's harbour and ferry pier. The 'Pettycur' milestone outside the ferry building gives the distance to Newport as 0.

Drive under the road bridge and follow signs to the carpark. The 2km incline gives the bridge an exaggerated perspective as seen from the 'aeroplane-wing' monument. The bridge leads right into the city of Dundee but if you think 'Great, no toll!' going in, be prepared for a double toll on leaving. *Discovery Point* and other features have given the town a new tourist interest.

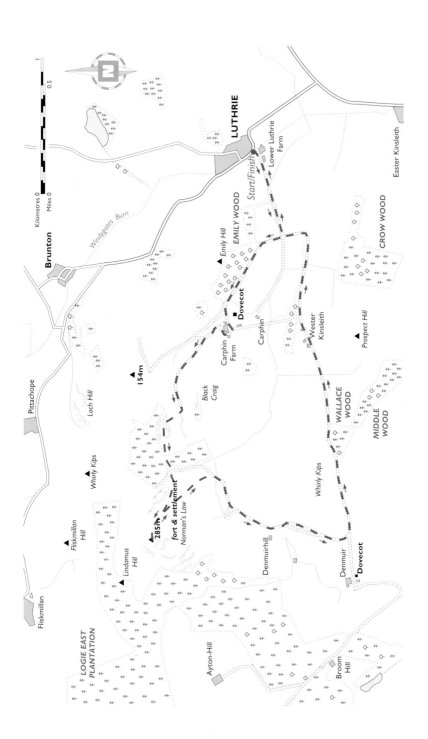

NORMAN'S LAW

INFORMATION

Distance: 8km (5 miles).

Start and finish: Luthrie village, off A914 (Kirkcaldy–Tay Bridge road).

Terrain: Mostly roads, tracks and paths but rough, open ground on the hill itself. Strong footwear advised.

Toilets/refreshments: Cupar, 5 miles; Newburgh, 8 miles.

Please note: Bulls and cows with young calves can be aggressive and their fields are to be avoided. Dogs should NOT be taken on this walk. All farm notices should be heeded.

Park carefully in the village, and set off past Lower Luthrie farm (footpath sign for Ayton) on a long, straight road. At the end of this straight, turn right to go up past a house and along outside Emily Wood. The path then dips and rises to Carphin Farm. In the field, left, is an unusually slender lectern doocot. Swing on up through Carphin and, after a levelling-off, take a left-fork track to climb steeply again (gate). Rolling green fields, hotching with rabbits and with rooks calling from the Scots pines, lie over to the right.

The track keeps swinging left, but – on nearing an obvious col – break off to climb the hill on the right. There is a sudden feeling of height, for underfoot are blaeberry, heather, tormentil, bedstraw, milkwort . . . and the summit is bare rock while the view is wide with the first sight of the Sidlaws, the Tay estuary and Dundee. Norman's Law looms ahead.

Head towards Norman's Law but then go down the left flank of the hill to avoid some cragginess and on along outside the fenced area to a stile, which gives access to the final slopes. There's another subsidiary bump to go over before Norman's Law will be gained. There are Iron Age

Brunton and Green Crag.

The thin doocot at Carphin Farm on the way up Norman's Law.

encampments and a fort on top, and the defences of the latter are very clear. The situation is fantastic. 'The view's better than from the top of Ben Nevis,' I've heard proclaimed – and wouldn't argue.

There are Highland hills looming, from Ben Lomond to Lochnagar (Lawers, Schiehallion, Farragon, Beinn a' Ghlo) and a view indicator will help pick them out. There's also a trig point and a large cairn. The view upriver is exquisite and varied, and in early summer it is patched with the bright yellow of rape fields and the dark green of conifer plantations. This is a summit for a late-evening visit to catch the sun dipping westwards.

The names around here are intriguing. On the Tay you would expect a Dog Bank but what of Eppie's Taes Bank, or Sure-As-Death Bank or Peesweep Bank? There are cheery land names: Robin's Brae, Cherrybank, Sunny Den and Blinkbonny; awful names: Dark Law, Foodie, Gallow Hill; amusing names: Lindifferon, Glenduckie; and try getting your tongue round Mountquhanie or Cunnoquhie.

Start off down by the way ascended as far as the first dip then peel off right, heading south-east and keeping close under all the rockiness of the minor bump to meet a fence and a gate (straight ahead) with a minor track. Go through the gate and follow this rutted, green way along then twisting down to a gate into a field whose edge is followed. The slopes of whin, right, can be heavily scented in May/June.

The track is clearer after the stile out of the other end of the field and angles along on the flank of Whirly Kips. Denmuir, below, is a large farm and has a big lectern doocot visible. The track finally swings left, bringing the often-seen pencil of the Hopetoun Monument dead ahead. The path goes over a stile then curls off down to Denmuir. You, however, turn left on a less-used track but one which long predated the busy A-roads of today down on what, in those days, was the boggy low ground. Robert Baillie of Luthrie created many of the 18th-century roads hereabouts.

You simply follow this old route from Ayton and Denmuir over to Luthrie, at first up a near-tunnel of thorn, elder, nettles, wild roses and gean then with fields on the left allowing good views up to Norman's Law again then as a lane to Wester Kinsleith and, finally, a firmer avenue down to the start, thus completing a memorable round. But do keep Norman's Law for a clear day. Cattle and sheep graze in many places so dogs should not be taken, and young children could find it too hard.

Denmuir and its doocot from below Norman's Law

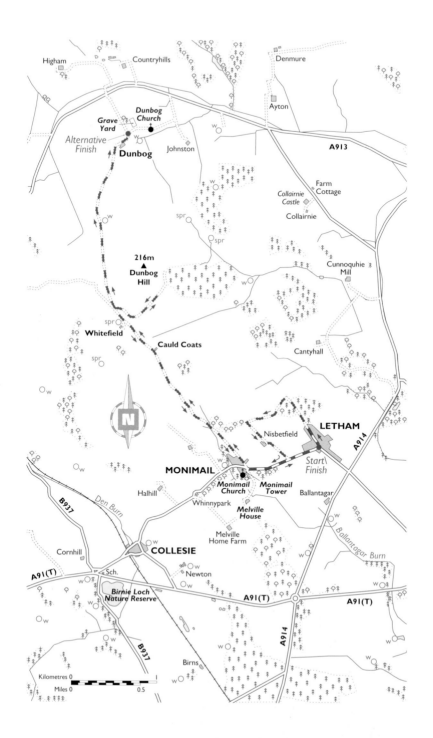

LETHAM, MONIMAIL AND DUNBOG

This is another corner of Fife that reflects times past, taking in the hamlets below the hills north of the Howe of Fife; yet, seeing the prosperous modern farms, the past is almost inconceivable. At the start of this century Letham had a shoemaker, blacksmith, saddler, cartwright, joiner, tailor, baker, grocer and butcher, there was a malt barn and every cottage had its loom.

Letham is a common east coast name so make sure you find the right one! Turning off the A914, park somewhere on the street of little cottages. At the top, on the west side, is the surviving village pump. Set off up School Brae, but first go to the far end of the row of houses, right of the brae, and keek round the back to see a charming pepperpot doocot (private garden, so don't go in, please). Halfway up School Brae there is a lectern doocot, right, and, where the road forks, if you walk leftwards, you'll see another one. Return and continue on the track up into the hills.

At the end of the trees, fork left round the wood to traverse the south-facing hillside with its fine view. Close all gates. Turn left on the farm track down to Nisbetfield. Monimail church is seen over its roofs and is your next port of call, an elegant whinstone building (1794) with a later tower. Clear windows allow a look inside. Walking on, you come to Monimail, a place that seems about to be swallowed

INFORMATION

Distance: 7km (4½ miles) to Dunbog; 9km (5½ miles) returning to Letham.

Start and finish: Letham (7km west of Cupar, on a minor road between the A914 and A91/B937 junction).

Terrain: Good tracks, minor road.

Toilets/Refreshments: Nearest are in Fife Animal Park or in Cupar and Newburgh.

Opening hours: *Fife Animal Park:* daily, 1000–1700; later in summer.

Letham.

Monimail's rich rural landscape.

by jungle. In summer the scent of ransoms hangs on the air. The churchyard is worth a visit, with war memorial gates and little stories on the stones. There are plenty of toffs' mausoleums, including the large Melville aisle of the original church. Another is full of Balfours of Fernie Castle and, outside the entrance, a Balfour who was the local blacksmith for 50 years.

Turn right from the gates to go through a hidden door to view the Monimail Tower, best known as part of the palace home of Cardinal Beaton, murdered at St Andrews in 1546. His successor as archbishop, Hamilton, was hanged in 1571 and, eventually, the estate was acquired by the Melvilles, who abandoned the tower to build sumptuous Melville House. If you walk down the drive from the immaculate North Lodge you can glimpse the mansion (now a school) then turn back and cross the road to head north into or over the Dunbog Hills. Monimail House is the old manse, and you pass the new graveyard. Monimail Tower can be visited (small exhibition) while restoration continues on it and other features in the walled garden.

The track follows a long valley. The farmstead of Whitefield, which sits on the col at its head, has been restored. Shortly before it, the ruins on the right are of Cauld Coats, a name appearing on a 1775 map, which shows the way from Money Meal [sic] to Dunbog but not the modern road. As the track turns to its end at Whitefield, go through a gate and straight on for one of those instant views that are unforgettable. You could turn back from here, but it is worth while going to the next gate and through it to the good track beyond.

Turn right up this track for a few minutes only, simply to see a view in the other direction to the pillar of the Hopetoun Monument – usually over oilseed rape fields and perhaps with skylarks paragliding overhead. Don't go to the end or up Dunbog Hill, but either head back to the car via Monimail and the minor road or, if you have arranged for someone to pick you up by car, take the track right round and down to Dunbog, which is actually the shorter option.

Descending north there's an unusual view through the Lindores gap to the waters of the River Tay, and Dunbog church occupies a wide east–west vale so is very prominent. It is now a private house. The graveyard's main interest is the remnant of its watch house with the inscription: '1822 Erected for protecting the dead.' If you are being met at Dunbog, both Newburgh and Cupar are reasonably handy for refreshments.

If returning to Letham drive west, through Monimail again, to see Collessie, another hamlet of character with a dominant church and one row of weavers' cottages restored as a house with a thatched roof. Between Monimail and Collessie the road does a wiggle at a cottage. This belonged to a weaver whom the Earl of Melville tried to remove in order to straighten the road. The cottage – and the wiggle – remain.

Beyond Collessie there's a double junction, and if you turn left on to the A91 (Cupar direction) and, almost at once, turn right, you will find Birnie Loch Nature Reserve on the left. The owner of a quarry created this after the site had been worked out, and it was given to the local council in 1992. Already over 50 species of birds have been noted (including ospreys). Some 300m on is the Fife Animal Park, a children's joy, which has a tearoom.

Fernie Castle near Letham.

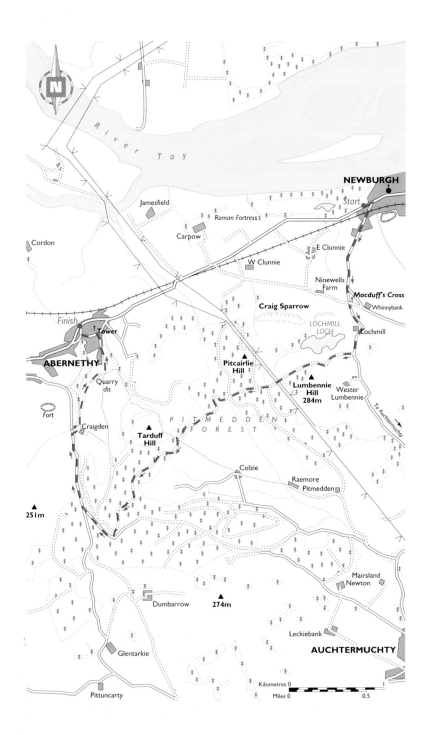

NEWBURGH TO ABERNETHY BY PITMEDDEN FOREST

Newburgh ('new' in 1266!) and Abernethy are both interesting old towns squeezed in between the rump of the Ochils and the big River Tay, and both deserve some exploration. Every house along Newburgh's High Street seems to be attractive and the whole is something of an 18th-century showpiece.

The town came into being in connection with Lindores Abbey (1178), the scant red ruins of which lie at the east end of the town. It had many royal connections and a bloody history but the records have nice touches, such as the monks having Papal blessing to wear bonnets because of the climate and their constant feud against adders. They introduced fruit growing to the area. The closure (and demolition) of the linoleum works has left the town rather lifeless. Newburgh falls down to the river opposite huge reed beds, some of which are used for thatching houses, another near-extinct craft. At the western edge of the town is a large carpark, next to the bowling green and war memorial. The walk starts up Woodriffe Road, opposite the war memorial.

The road climbs steeply and opens out a grand vista over the Earn and Tay, a view that had Sir Walter Scott rating it world class. Keep right when the road forks and, at cross tracks (Ninewells Farm entrance) turn left to see the historic MacDuff's Cross, a cup-marked sandstone block, to which many legends are attached.

Continuing, there's a glimpse through to Lindores Loch. The road steepens and twists up by Lochmill Farm, with the wall of a dam above it, and when it tops the crest there is a track going off right, which is your route. There are views down to Lochmill Reservoir as the track climbs along by Craigdownie and the forest edge of Lumbennie Hill. Power lines cross the route and buzzards may be heard mewing overhead. At a junction turn right. Having been felled and replanted there are better views down

INFORMATION

Distance: 9km (6 miles).

Start: Newburgh, on the River Tay, east of Perth. Carpark at west end.

Finish: Abernethy. A bus service operates to Newburgh.

Terrain: Road, forest track and good footpath.

Toilets/refreshments: Newburgh and Abernethy.

Opening hours: *Abernethy Round Tower:* weekdays, 1000–1700; Sunday, 1200–1700, key from café opposite. *Laing Museum, Newburgh:* April–September, Monday–Friday, 1100–1700, Saturday–Sunday, 1400–1700; October–March, Friday, 1200–1600, Sunday, 1400–1700.

The Pictish symbol stone – Abernethy.

Lochmill Loch.

on the Pitmedden Forest side. Drop down to the evocatively named Seven Gates, a multiple junction.

Head straight on, twisting upwards, and enjoy the magnificent views into Fife. Ironically, you are in Perthshire here, with what many regard as the finest view of the Lomonds of Fife, while earlier, at Scott's View (MacDuff's Cross), you were in Fife but the view was of Perthshire. It's ironic, too, that this large area of conifer planting is called Pitmedden Forest when the Pitmedden hollow below you (with its yellow collar of whin) is about the only area not afforested. The forest closes in for a while then you begin to lose height; on the right there's a greened-over lochan then you come to the edge of the old forest again, which is joined by a track from Newhill and Reedie Hill (red deer farm). Outside this forest is the young Glen Tarkie forest covering Dumbarrow Hill. Descend steadily to Abernethy Glen and join the road from Perth into Fife (Strathmiglo), turning right for 1km to come to Craigden. Watch out for speeding cars on this twisting road.

At the entrance to Craigden, turn left for a

footpath down Abernethy Glen. This goes by the Ballo Burn then steps lead up out of the 'den' on to the Witches Road, a traversing path named after a real enough 'coven', who were judicially burned and their ashes scattered on Abernethy Hill in the 17th century. Looking across Abernethy Glen there is the prominent bump of Castle Hill with an old quarry on its slopes. There's a prehistoric fort on the prow. Its name, Preaching How, perhaps refers to Covenanting times and Quarrel Knowe indicates where archery once took place. You come out at Loanhead Quarry, and the track becomes Kirk Wynd as it leads down to the square.

Abernethy's landmark feature is its 'Irish' Round Tower, dating to the 9th century, and, with the one in Brechin, the only survivor of its type in Scotland. Unlike most Round Towers this one can be climbed. The churchyard is also worth a browse and, on the tower, note the jougs (collars for malefactors) and an old Pictish symbol stone. Earlier, the Romans had a ferry at Carpow.

Abernethy round tower – one of only two 'Irish' towers left in Scotland.

The first church here was c. AD 460, and the Pictish kings made Abernethy their capital. Only the tower remains to indicate any glorious past. Turn west for the main road, the A913, where there is a bus stop opposite the one-time coaching inn (Abernethy Hotel). You pass Tootie House, its name a reminder of how a herdsman would blow a horn each day to remind people that it was time to take their cattle to the common grazings.

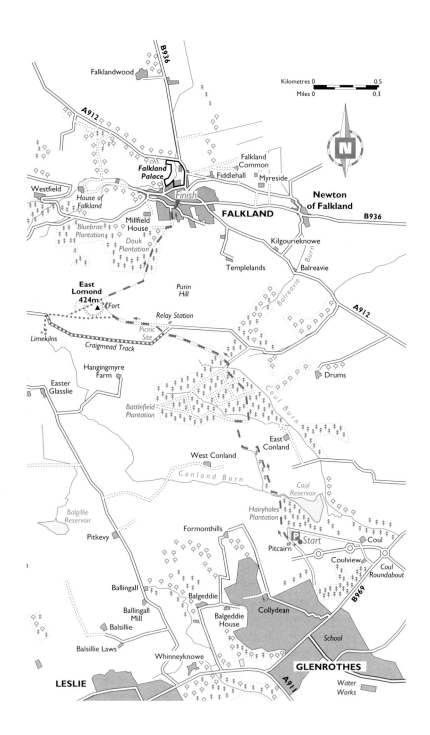

Kilometres 0   0.5
Miles 0   0.3

N

B936

Falklandwood

A912

Falkland Common

Falkland Palace

Fiddlehall

Myreside

Westfield

House of Falkland

Finish

FALKLAND

Newton of Falkland

B936

Millfield House

Bluebrae Plantation

Douk Plantation

Kilgourieknowe

Balreavie Burn

Templelands

Balreavie

A912

East Lomond 424m

Fort

Purin Hill

Relay Station

Picnic Site

Limekilns

Craigmead Track

Coul Burn

Drums

Hangingmyre Farm

Easter Glasslie

Battlefield Plantation

East Conland

West Conland

Conland Burn

Coul Reservoir

Balgillie Reservoir

Hairyholes Plantation

Pitkevy

Formonthills

P *Start*

Coul

Pitcairn

Coulview

Coul Roundabout

Ballingall

Balgeddie

Collydean

B969

Ballingall Mill

Balsillie

Balgeddie House

Balsillie Laws

Whinneyknowe

GLENROTHES

School

LESLIE

A911

Water Works

EAST LOMOND
AND FALKLAND

This walk crosses the Lomonds *and* incorporates a stroll in historic Falkland. Being a traverse route a second car must be left, or a willing driver found, to drop and pick up the walkers. East Lomond (Falkland Hill), at 424m, is the second-highest point in Fife and has a sweeping view from the summit.

Glenrothes sprawls up the lower slopes of the Lomonds, so follow directions carefully to the start from the main road, skirting the east side of the Lomonds (A912, A92). Turn off on the B969 and at the first roundabout (Coul Roundabout) turn right for Pitcairn. Coming from the west (Leslie) turn left at the first roundabout on to the B969 and the Coul Roundabout is the second roundabout thereafter. Turn left for Pitcairn. (A pseudo stone circle decorates the roundabout.)

Go straight on at the next roundabout then, at a second, turn right into an area being developed, where a sign indicates a small road, left, for Pitcairn and leads up to a carpark. The cottages are the Ranger Base and the start has a notice-board and map.

This is a Public Footpath Agreement route in the Fife Regional Park, so it tends to follow wood edges in angular regimentation, while power lines spoil early views. The first plantation bears the name Hairyholes, and there's a dip down and up to cross the Conland Burn. An open field leads up to the larger Battlefield Plantation and the route becomes a forest track for a while. This comes to a T-junction and, turning right, there is a view along the Howe of Fife. The path then heads left, up the short stretch of forest remaining, to gain the blaeberry/heather moors at a delightful beech clump. Rowans are sprouting as there is little grazing. An old wall is followed up to the carpark by the relay station (notice-board, picnic tables and toilets).

INFORMATION

Distance: 6km (4 miles) direct.

Start: Pitcairn Ranger Base, near Glenrothes.

Finish: Falkland (big carpark, signposted).

Terrain: Mostly footpaths or in town, but the steep hill calls for boots and fitness.

Toilets: Carpark at East Lomond and Falkland.

Refreshments: Wide choice in Falkland.

Opening hours:
Falkland Palace: April–October, Monday–Saturday, 1100–1750, Sunday, 1330–1730.
Town Hall: 1330–1730 only.

Less than 1½km west along the Craigmead track lies a restored limekiln with interpretative display boards that could be taken in on a half-hour diversion.

From the carpark cut through on to the track that runs along by the old wall for a while (willow scrub) then goes through a gate (warning sign to keep dogs on lead) for the final upthrust of the hill, which is a remnant of a volcanic outpouring. Oddly, a trig point lies off left on a lower shelf; the bare, flat summit only has a view indicator.

As you tackle the final steep slope, Falkland appears, down right, with a path leading clearly to the plantation above it. This is your route down, and to prevent erosion on the direct descent (which is dangerously steep anyway), please return here to head down at this easier angle, joining the path at the gate through a fence.

The summit view takes in Highlands, Lowlands and the Southern Uplands, as well as the sweeping Forth Estuary. The rest of the Lomonds are close, and below lies the restored limekiln. Beyond Craigmead the central bump among the bumps is Maiden Castle, a prehistoric fort, as, indeed, is the summit. (Down on the Falkland side the circling walls are still clear.) Norman's Law (and the Hopetoun Monument) and Largo Law are other walks clearly seen.

Head down as indicated above. The path from the gate wanders down the moor to reach the Douk Plantation. After this you will be glad to leave the crypt-like gloom of conifers into a cheery choir of beech wood. An opening

One of the historic inscriptions in Falkland.

gives a view to Falkland's spires and palace and 'the works'. A steep slope leads down to a driveway; turn right and down into Falkland. The St John's Works began as a weaving factory but 60 years ago it changed to making linoleum. Now the works produces plastic bags. Turn right just past the works and when the road swings left, you are suddenly in 'old' Falkland. Pause at the first crossroads.

The house on the right has a marriage lintel dated 1663 (the earliest in the town is 1610), and, on the left, note the attractive Horsemarket, with features like forestairs, harling, crowstep gables, pantiles and dated lintels. Further on, facing the Stables Gallery (corner, right), is a hostel, which may originally have housed the children brought in to slave in the weaving factory.

The street leads out to the central High Street – East Port, with town hall, palace and fountain. Before exploring the palace look at the plaques on the houses opposite (among them is the Hunting Lodge Hotel) – it is probably worth buying a local guidebook for the palace and town. The Royal Palace (NTS) is the major attraction, the 16th-century gatehouse and elegant Renaissance frontage dominating the town centre. It was the much-loved hunting seat of the Stewart monarchs, somewhere to escape the burden of state duty – a sort of early Balmoral. Its highlights are the Chapel Royal, the King's Room, the recreated gardens and the 1539 Royal Tennis Court, which is still in use.

Falkland Palace gateway.

The Bruce Fountain may be on the spot of the original town well. The heraldic lions hold shields displaying the Bruce and town coats of arms, the latter a stag lying below an oak. The nearby statue is to Onesiphorus Tyndall-Bruce, the local laird, who died in 1855. The shape of a cross on the road indicates the site of the town's mercat cross.

Walk up Cross Wynd and turn right along Brunton Street. The coat of arms over one door is of the hereditary royal falconer. At the end, turn sharp right along the High Street back to the town centre, passing the improbably sited local electricity sub-station. Cameron House was the likely birthplace of the famous covenanter whose name was adopted by the now-disbanded Cameronians regiment.

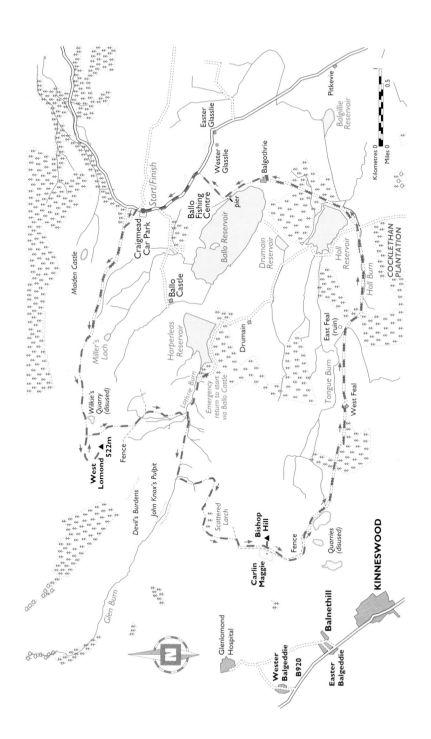

Pitkevie

Balgillie Reservoir

Easter Glasslie

Wester □ Glasslie

■ Balgothrie

pier

Start/Finish

Ballo Fishing Centre

Ballo Reservoir

Drumain Reservoir

Holl Reservoir

Craigmead Car Park

Maiden Castle

■ Ballo Castle

Miller's Loch

Harperleas Reservoir

Drumain □

Holl Burn

COCKLETHAN PLANTATION

Wilkie's Quarry (disused)

Lothrie Burn

Emergency return to start via Ballo Castle

East Feal (ruin)

West Lomond ▲ 522m

Fence

Tongue Burn

West Feal

Devil's Burdens

John Knox's Pulpit

Scattered Larch

Bishop Hill ▲

Fence

Quarries (disused)

Carlin Maggie ▲

KINNESSWOOD

Glen Burn

N

Glenlomond Hospital

Wester Balgeddie

B920

Balnethill

Easter Balgeddie

Kilometres 0
Miles 0
0.5

A LOMOND HILLS CIRCUIT

T his is a major hill walk and, taking in the highest summit in Fife (West Lomond, 522m), it is worth saving for a clear day. It should be treated as a proper mountain expedition: you must go well equipped and supplied. (In summer carry water.)

The Lomonds loom large in views of Fife (though, to be accurate, the Bishop is in Kinross-shire), and the high road from Leslie to Falkland, which divides the range, gives easy access and considerable height before starting: Craigmead carpark is at 285m. From Falkland you simply drive along the High Street and the continuation is indicated. From the south (Glenrothes/Leslie) some care is needed as the turn-off in Leslie (a narrow road near the east end of the town) is difficult to see, though signposted for Lomond Hills/Falkland. Craigmead is obvious and has toilets, information/map board and picnic tables.

The walk starts in the corner of the carpark – signed for West Lomond – runs along a hollow and up steps. Turn left on the green paths that soon merge with a track coming up from the edge of a plantation and follow the track over the moorland. (The Maiden Castle, a prehistoric fort, with very clear defensive ditches at its base, is well worth the detour for those interested in ancient sites.)

The big track gives easy walking. There is one stile and then, at the foot of the final cone, you must take the slope off around the hill to the right, a diversion that is slowly healing the scar of the path that once ran straight up to the summit. The path rises steadily and reels in an ever-changing view: Highlands, Ochils, far Ben Vorlich and Stuc a' Chroin, Loch Leven, the Cleish Hills . . . before finally reaching the white trig point and the black stones of a cairn. 'Brilliant!' was the response of one youngster.

Retrace the ascent route to the diversion notice at the

INFORMATION

Distance: 17½km (11 miles).

Start and finish: Craigmead carpark on the hill road over the Lomonds from Leslie to Falkland.

Terrain: Mainly on good paths and tracks but some rough hillsides, so the walk should be treated as a mountain expedition, and you should wear boots and carry the appropriate gear. Take food and water.

Toilets: Craigmead carpark.

Refreshments: None en route; Falkland afterwards (3km).

Glen Vale and John Knox's pulpit.

foot of the final cone and continue round the foot of the hill, more or less south then south-south-east, aiming eventually for some 'crumpled' ground, where several springs and streams start. There is a fence, anyway, and this is crossed at a stile, the Three Marches stile, just east of the broken ground. Descend the steep slope of the field – near the deep cut of the burn – and when the level ground is reached, cross a stile and continue along the other side of the wall. There's a stile in the corner, which is on the ancient route from Glen Vale to Harperleas, and the path swings west to join this old path. (The infant Lothrie Burn could be the only water for drinking for the rest of the walk.) If anything demands a shortening of the route, you can regain the start by going east via Harperleas Reservoir and the ruined Ballo Castle.

This wide col lies between West Lomond with its weathered exposures (the Devil's Burdens) and the wide flank of the Bishop. The track west is good and when it swings over to the wall (left) there is a gate, which you go through, but first continue for a few minutes to look down Glen Vale, a place worth exploring on another walk.

Once through the gate, follow the track up the slope until a gate in the wall line is reached. Just before it, turn right, through the broken walling, and head west up into a landscape which looks like a carpet that has been rumpled into ridges and hollows, through which you pick your way (good sheep paths), and being surprised, too, by discovering a scattering of mature larch trees in this unlikely spot. Work up and leftish, till the path/overgrown wall along the scarp of the hill is reached. It is a dizzy drop on the other side, and the prevailing west winds, hitting this barrier of hill, shoot up to give excellent lift to the gliders based at Portmoak (Loch Leven). You may well see and hear them overhead and below.

Carlin Maggie, Bishop Hill.

After crossing a fence bear right to keep by the fence line as this gives a view of a pinnacle below the crags called Carlin Maggie (Carlin = witch). Tradition reports that Maggie had words with the Devil who was building West

Lomond and he dropped his burden and rounded on Maggie, turning her to stone. The slim 'figure' was taller but some years ago lost her 'head'. The pull beyond brings the summit bump of Bishop Hill (461 m) into view, a bit off-left, and marked by a cairn. Some would claim it as the best summit view in the Lomonds.

Return to the path along the fence on the scarp's edge. Pick out the path before leaving the cairn. It runs to a gate; you go through this and, beyond, the path becomes a green track, joined by another, which comes right up that steep west flank. It is worth peering over. The whole area hereabouts was once extensively quarried for limestone, and the track wends along to pass several defunct quarry sites before reaching the larch plantation and the track down to West Feal. In 1852 there was a 'gold rush' with thousands camped here and the villages below were cleared out of all supplies.

Timber extraction has rutted the old track badly but you can walk on the grass beside it. West Feal is no longer inhabited and East Feal, down the straight bit of track, is ruinous. A burn accompanies the road to some farm buildings and then you come to a crossroads and the tarred road that gives access to this hidden heart of the Lomonds.

To Arnot Reservoir from West Feal.

Cross over to drop down below the dam of the Holl Reservoir and past the Italianesque-looking water-treatment works. Beyond, after admiring West Lomond over the loch, go through a kissing gate and up the break in the plantation (also followed by a line of power poles). The crest above suddenly opens the view on to Ballo Reservoir, the largest of the six in the Lomonds.

Descend to Balgothrie Farm, where diversionary arrows should be followed to avoid the farm buildings. Continue beyond and, when the Ballo Fishing Centre complex is reached, keep to the field edge behind it then, just beyond, turn up along the field edge and keep on climbing till a last, large stile deposits you on the Leslie–Falkland road. Turn left and a 15-minute tramp will lead back to the Craigmead carpark.

INDEX